Prentice Hall International Series in Acoustics, Speech and Signal Processing

Programming Real-time Multicomputers for Signal Processing

Urban A. Thoeni

Prentice Hall
New York   London   Toronto   Sydney   Tokyo   Singapore

First published 1994 by
Prentice Hall International (UK) Limited
Campus 400, Maylands Avenue
Hemel Hempstead
Hertfordshire, HP2 7EZ
A division of
Simon & Schuster International Group

© Prentice Hall International (UK) Limited 1994

All rights reserved. No part of this publication may be reproduced,
stored in a retrieval system, or transmitted, in any form, or by any
means, electronic, mechanical, photocopying, recording or otherwise,
without prior permission, in writing, from the publisher.
For permission within the United States of America
contact Prentice Hall Inc., Englewood Cliffs, NJ 07632

Printed and bound in Great Britain by
Redwood Books, Trowbridge, Wiltshire

Library of Congress Cataloging-in-Publication Data

Thoeni, Urban.
    Programming real-time multicomputers for signal processing / Urban
Thoeni
      p. cm.
    Includes bibliographical references and index.
    ISBN 0-13-489857-5 (pbk.)
    1. Signal processing—Data processing. 2. Real-time programming.
3. Microcomputers—Programming. 4. Parallel processing (Electronic
computers) 5. Multiprocessors.   I. Title.
TK5102.9.T46 1994
621.382'2'028552—dc20   93-40767
                                                                              CIP

British Library Cataloguing in Publication Data

A catalogue record for this book is available from
the British Library

ISBN 0-13-489857-5

1 2 3 4 5   98 97 96 95 94

To those who love and encourage me:
my parents Claudia and Gion Peder
my wife Angelika
our children Severin, Roman and Ursina

Contents

Preface   xi

Nomenclature   xiii

1 Introduction
1.1 Real-time Signal Processing and Data Flow Computations   1
1.2 Book Outline   3

2 Problem Outline   5
2.1 Types of Multiple-processor Systems   5
2.2 Real-time Systems   7
2.3 Aims of Optimization   8
2.4 Trade-offs between Aims and Feasibility   9
2.5 The SISAL and OCCAM Programming Languages   10
2.6 Data Flow Graphs   12
2.7 Survey of the Processing Steps for Partitioning and Allocating an Application Program   13
2.8 References   15

3 Survey: Algorithms for Control and Signal Processing and their Characteristics   19
3.1 Introduction   19
3.2 Notation for the Description of Systems   20
    3.2.1 Linear Systems   20
    3.2.2 Nonlinear Systems   21
3.3 Linear Controllers   21
    3.3.1 Single-input Single-output (SISO) Controllers   21
         The PID Controller   22
         General Dynamic SISO Controller   23
    3.3.2 Multiple-input Multiple-output (MIMO) Controllers   23

		Linear State Feedback	24
		Linear State Feedback with Estimator	24
		Optimal Estimator: The Kalman Filter	24
	3.4	General Systems	25
		3.4.1 Linear Systems	25
		3.4.2 Nonlinear Systems	25
	3.5	Solving Nonlinear System Equations by Numerical Integration	26
		3.5.1 One-step Method: Fourth-order Runge-Kutta Algorithm	27
		3.5.2 Multistep Method: Fourth-order Adams-Bashforth Algorithm	28
	3.6	Signal Processing Applications	29
		3.6.1 Digital Filtering	30
		Finite Impulse Response (FIR) Filters	30
		Infinite Impulse Response (IIR) Filters	31
		3.6.2 Signal Transforms	32
		Discrete and Fast Fourier Transforms	32
		Recursive Formulation of the One-dimensional Mixed-radix FFT	33
		Matrix Formulation of the FFT	37
		Convolution, Correlation, and Spectral Estimation	40
	3.7	Matrix Computations	41
		3.7.1 Basic Vector and Matrix Computations	42
		Scalar-vector Computations	43
		Vector-vector Computations	43
		Matrix-vector Computations	44
		Scalar-matrix Computations	45
		Matrix-matrix Computations	45
		Block Matrix Computations	46
		3.7.2 Solution of Systems of Linear Equations	46
		3.7.3 Matrix Inversion	47
	3.8	References	48
4	**Preparing the Data Flow Graph for Partitioning**		**53**
	4.1	Data Flow Graph Generation	53
	4.2	Communication Volume Analysis	56
	4.3	Graph Expansion	56
		4.3.1 Forall Nodes	57
		4.3.2 Function Calls	58
	4.4	Execution Cost Analysis	59
		4.4.1 Communication Costs	59
		4.4.2 Node Execution Costs	61
	4.5	References	62
5	**Partitioning the Data Flow Graph into Tasks**		**63**
	5.1	Building Tasks	63

	5.2	Approaches Described in the Literature	65
	5.3	Deadlock Avoidance	68
	5.4	Partitioning the Data Flow Graph	72
	5.4.1	*Rules For Building Tasks*	72
	5.4.2	*Preparation of the Data Flow Graph*	74
	5.4.3	*Simulation of the Data Flow Graph's Execution*	76
	5.4.4	*Computational Complexity Analysis of Partitioning*	77
	5.4.5	*Summary of the Properties of the Partitioning Technique*	78
	5.5	References	79
6	**Static Task Allocation and Code Generation**		**81**
	6.1	Dynamic versus Static Scheduling and Task Allocation	81
	6.2	The Static Task Allocation Problem	84
	6.2.1	*General Machine Configuration*	84
	6.2.2	*General Static Task Allocation Problem*	85
	6.2.3	*Static Task Allocation Problem for Homogeneous Real-time Multicomputers*	88
	6.3	Solutions Proposed in the Literature	93
	6.3.1	*Graph Theoretic Approach*	94
	6.3.2	*Numerical Optimization Approach*	94
	6.3.3	*List Scheduling Approach*	95
	6.3.4	*Simulated Annealing Approach*	96
	6.3.5	*Clustering Approach*	97
	6.4	Two-phase Linear Clustering Approach	99
	6.4.1	*Reducing the Number of Interconnections*	101
	6.4.2	*Phase One: Clustering Heavily Communicating Tasks*	103
	6.4.3	*Phase Two: Matching Interconnection Topology and Communication Resources*	103
	6.4.4	*Computational Complexity Analysis of the Allocation*	104
	6.5	Translating the Partitioned and Allocated Data Flow Graph into Target Code	106
	6.6	Generating OCCAM Code	106
	6.8	References	108
7	**Case Studies**		**113**
	7.1	A Small Example	113
	7.2	Digital Filters	121
	7.2.1	*FIR Filters*	121
	7.2.2	*IIR Filters*	126
	7.3	State Space Controllers with Observers	131
	7.4	Nonlinear Controllers	134
	7.5	Numerical Integration	137
	7.6	Fast Fourier Transform	138
	7.7	Conclusions	143

8 Conclusions **145**
    8.1   Assessment of the Data Flow Approach for Real-time Systems   145
    8.2   Future Work   145
    8.3   Reference   147

Appendix A: Specification of the Multicomputer **149**
    A.1   Multi-Transputer System   149
    A.2   High-speed Data Acquisition System   150
    A.3   Firmware   152
    A.4   References   152

Appendix B: Detailed Specification of the Processing Steps for Partitioning and Allocating **153**
    B.1   Communication Analysis   153
    B.2   Graph Expansion   155
    B.3   Execution Cost Analysis   160
    B.4   Graph Partitioning   166
    B.5   Task Allocation   169
    B.6   OCCAM Code Generation   173

Index **181**

Preface

Parallel computing has been receiving much attention during the past years. New generations of ever more powerful computers are arriving at an increasing rate. While vector supercomputers are well understood and routinely used by scientists in all fields, the situation is different with multicomputers consisting of more than only a few processors. While hardware design has made rapid progress, the science of programming such computers is still lagging years behind. Therefore the use of parallel computers is confined mostly to specialized scientists and engineers.

There have been attempts to parallelize automatically computations formulated in conventional programming languages. However, such an approach resembles estimating the original size of the potatoes by inspecting the French fries. Since the computational concept behind the conventional programming languages is inadequate for expressing parallelism, new models have to be found. One of these approaches is the concept of data flow. No artificial serial order is imposed on the computations, but each operation is ready to execute as soon as its input values are available.

Whereas the efficient incorporation of the data flow approach into hardware has proved to be difficult, it is very useful for analysis of the interdependence of the computations. All the relevant information about the computations is preserved in the data flow graph representation. Particularly in message-passing multicomputers, where the data exchange contributes a substantial part to the program execution time, communication overhead can easily be used and minimized for program execution.

As an example of a functional language which allows a data flow analysis, SISAL (Streams and Iterations in a Single Assignment Language) and the associated data flow graphs are introduced. This language, jointly developed by universities and research laboratories, is now promoted by the Lawrence Livermore National Laboratory and has been ported to major computing platforms.

Apart from communication minimization the sheer complexity of mapping the computations to several processors prohibits a manual attempt. For the real-time applications in the project described in this book it would prove impossible to meet all constraints when partitioning and distributing an application algorithm to a multicomputer. Therefore, automated tools are indispensable for the process of finding a good distribution of the application algorithms to the processors.

This book describes a research project on automatic parallelization of computations for real-time signal processing and control. It was conducted at the Swiss Federal Institute of Technology (also known under the German acronym ETH) in Zurich. When I started my doctoral research project in electrical engineering on the fascinating area of high-performance parallel computing I quickly discovered that the basic question of how to distribute the computations to several processors was unsolved and I devoted myself to this problem. While the final and most general solution has not yet been found, the approach described in this book shows a viable method to solve the partitioning and allocation problem.

The target system has the architecture of a message-passing multicomputer. The aim of the allocation is to minimize the iteration time for a set of computations which is executed periodically in a non-overlapping manner. The execution and communication cost models are developed for the INMOS Transputer. This processor is easily available at a low cost to study multiprocessor systems. An additional advantage is the availability of the OCCAM parallel programming language which is specially designed to match the capabilities of the processors. (In fact, the development sequence was the other way around, since the processors were tailored to allow an optimum implementation of the language.) However, the cost models and the methodology of distributing the applications to the processors are independent of the Transputer and are applicable to all message-passing multicomputers.

Extensive literature surveys and references are given on the subjects of partitioning algorithms and data flow graphs as well as on the allocation of communicating tasks on multiple processors. The characteristics of the algorithms for signal processing and control are discussed and again all original references are given for additional studies.

I am indebted to many people who have contributed to the project described in this book. They include: Oskar Brachs, Hansjörg Diethelm, Monique Faber, John Feo and his SISAL group at the Lawrence Livermore National Laboratory, Hans P. Geering, Albert Kündig, Srdjan Mitrovic, Brigitte Rohrbach, Daniel Schweizer, Renato Zanetti, and the staff of the ETH library. The partial funding of the project by the Swiss National Science Foundation under the grant no. 21-26,648.89 is acknowledged.

This book is aimed at advanced students, scientists, and engineers working in the area of high-performance real-time and parallel computing. Some basic concepts of graph theory and computer science will be helpful, but need not exceed general engineering knowledge. The basics of signal processing and control algorithms are carefully developed in this book, followed by a detailed introduction to the architecture of general multicomputers and partitioning and allocation problems, together with the special problems posed by real-time environments.

I hope that some readers will be attracted by the fascinating problems connected with parallel computing and will perhaps become motivated to conduct their own research to gain additional insights into this developing subject.

Urban A. Thoeni

Nomenclature

A	system matrix, continuous or discrete controller, dimension n × n
	task assignment matrix, dimension m × p
A_{opt}	optimum assignment of the tasks to the processors
a	coefficients of the denominator polynomial of a transfer function,
	coefficients of the Runge-Kutta integration algorithm
B	input matrix, continuous or discrete controller, dimension n × m
	resource demand matrix, dimension m × p
b	coefficients of the numerator polynomial of a transfer function, coefficients of the Runge-Kutta integration algorithm, number of bytes transmitted
C	output matrix, continuous or discrete controller, dimension p × n, basic transform matrix (signal transform),
	resource capacity matrix, dimension m × p
c	index denoting the controller, coefficients of the Runge-Kutta integration algorithm
c_{int}	communication time for internal communication
c_{ext}	communication time for external communication
D	direct transmission matrix, continuous or discrete controller, dimension p × m
	overlap of periodograms for spectral estimation,
	communication distance matrix, dimension p × p
d	desired output of the control system
E	twiddle factor multiplication matrix (signal transform)
$E(T_k)$	execution costs of task k
e	control error, difference of desired and actual output of the plant
F	system matrix, continuous plant, dimension n × n
$F(\Pi_i)$	execution cost of partition i of a data flow graph
F_c	communication cost of a partition of a data flow graph
f(.)	non-linear function in a system equation
G	input matrix, continuous plant, dimension n × m
G(s)	continuous transfer function

xiii

$G(z)$	discrete transfer function
H	output matrix, continuous plant, dimension $p \times n$
H_d	output matrix, discrete plant, dimension $p \times n$
h	integration step size, impulse response of a dynamic system
I	identity matrix
i	general counting variable
J	direct transmission matrix, continuous plant, dimension $p \times m$
J_d	direct transmission matrix, discrete plant, dimension $p \times m$
j	imaginary unit, $j = \sqrt{-1}$, general counting variable
K	control gain matrix, dimension $m \times n$
K_P, K_I, K_D	coefficients of the PID controller
k	index denoting the time instance for sampled systems, general counting variable
L	estimator gain matrix, dimension $n \times p$
	length of periodograms for spectral estimation
	link matrix, dimension $m \times m$
l	general counting variable
M	$E\{(x-\bar{x})(x-\bar{x})^T\}$, set of tasks to be allocated ($|M| = m$)
m	dimension of the control vector u, number of factors of the number of data points of a signal transform, number of tasks to be allocated
N	number of data points of a signal transform, number of nodes in a data flow graph
N_w	dimension of the plant input disturbance
n	dimension of the state vector, degree of the denominator of a transfer function, general counting variable
O	communication overhead cost matrix, dimension $m \times m$
$O(T_k)$	overhead cost of task k
P	$E\{(x-\hat{x})(x-\hat{x})^T\}$, permutation matrix (signal transforms), set of processors in a system ($|P| = p$)
PA^∞	task assignment with unbounded number of processors
p	dimension of the output measurement vector y, number of processors in a system
q	degree of the numerator of a transfer function
R	set of resources in a system ($|R| = r$)
R_v	variance of the plant output disturbance, dimension $p \times p$
R_w	variance of the plant input disturbance, dimension $N_w \times N_w$
r	factor of the number N of data points of a signal transform, correlation sequence, number of resources in a system
S_{xy}	cross power density spectrum of the signals x and y
$\hat{S}_{xy}$	estimated cross power density spectrum of the signals x and y
s	complex variable for continuous transfer functions and the Laplace transform
T	sampling time interval, transform matrix (signal transforms)
T_{setup}	communication setup time

T_{memI}	memory access time, internal communication
T_{memE}	memory access time, external communication
T_{trans}	message transmission time
T_{min}	minimum task execution cost
T	transpose symbol
t	continuous time
TP	task parallelism matrix, dimension p × p
U	set of processors in the final allocation ($|U| = u \leq p$)
u	input/control vector, dimension m × 1, load of a processor
$\bar{u}$	mean load of all processors
V	communication volume matrix, dimension m × m
v	plant output noise or disturbance, dimension p × 1
W	set of non-faulted processors ($|W| \leq p$)
w	plant input noise or disturbance, dimension N_w × 1, number of long words (4 bytes) addressed
w_i	allocation priority of task i
w_N	abbreviation for $e^{-\frac{2\pi j}{N}}$
x	state vector, dimension n × 1
$\hat{x}$	estimated state vector, dimension n × 1
$\bar{x}$	predicted state vector, dimension n × 1
y	output measurements, dimension p × 1
z	complex variable used for discrete transfer functions and the Z transform
α	factor for vector multiplication
β	coefficients of the Adams-Bashforth integration algorithm
$\partial(k)$	Kronecker function, $\partial(k) = 1$ if $k = 0$, else $\partial(k) = 0$
Φ	system matrix, discrete plant, dimension n × n
Γ	input matrix, discrete plant, dimension n × p
$Γ_1$	noise matrix, discrete plant, dimension n × N_w
λ	complex variable describing the poles of a system
$Π_i$	partition i of a data flow graph
σ	partitioning efficiency
*	complex conjugate

Time signals are denoted by lower-case letters and their corresponding transforms by capital letters.

CHAPTER 1
Introduction

1.1 Real-time Signal Processing and Data Flow Computations

Real-time signal processing and automatic control have gained much importance in the last years. New applications for medical systems, image processing, robotics, avionics, financial data systems, seismic data processing, intelligent control, and many more areas have appeared at a fast pace.

These developments have become possible mainly through the rapid development of computer technology. The computational power of the mainframe systems of some years ago is now delivered by desktop systems, and the capacity of mass storage devices has surpassed limits never imagined previously.

The exploitation of parallelism in computing systems has evolved from simply fetching one instruction and one operand from memory at the same time, or pipelined instruction execution in the central processing unit to much more complex operating principles. Now entire systems consisting of multiple highly pipelined vector supercomputers are cooperating to execute a program.

However, the developments of the hardware have left the software technology for programming the computers far behind. The programming language Fortran, developed in the 1950s, is still the programming language used most often for scientific applications. It is kept alive by regular updates, but is totally inadequate for modern computers due to its underlying concept of sequential execution on one processor.

Worse still, the formulation of computations in such a language often prevents an efficient utilization of novel computer architectures through the introduction of artificial dependencies among computations, through ambiguities (i.e., use of the same variable for completely different purposes) and uncontrollable sharing of memory regions (Fortran's COMMON blocks). Thus it becomes impossible to separate the relevant information about the computations to be performed from the artificial restrictions imposed by the means of representing the computations. So far, only a few and relatively crude tools have become available to aid the task of implementing computations on multiple processors.

In contrast, analysis of the flow of data through a graph representing a set of computations reveals the true dependencies of the atomic operations among themselves. No artificial serial order is imposed upon the computations as is the case when they are formulated in a conventional programming language.

However, implementing a real dataflow environment on a conventional computer is too costly from the point of view of task administration and synchronization overhead. Another crucial issue is fast communication at predictable costs.

Message-passing architectures as introduced by the Transputer family of computers have opened up new prospects since they handle communication and synchronization in special hardware units. The powerful central processing unit, together with a fast microcoded scheduler, provides an environment for producing systems of tasks where the execution of the processes is data-driven and handled efficiently. The only condition is that tasks must be allocated statically to the processing elements.

The algorithms used in signal processing and control possess the properties for execution on such an architecture. They can be formulated to use only static data, and the number of processes and their interdependencies can be determined in advance.

Through the self-synchronization of the computations by communication only minimum delays are introduced. If the communication is sufficiently fast then the minimum execution time of the computations is achieved by performing operations in parallel that are independent of each other.

This is where parallel processing comes into play. Given enough processors and an appropriate distribution of the computations to each of them, the advantages of the data flow graph formulation of an algorithm can be fully exploited. Even if fewer processing elements are available than the number of independent tasks, it is possible to distribute the computations in such a way that the time needed to perform them is significantly shorter than that required for purely serial execution.

If the execution time is reduced by utilizing several processors, then either the execution rate of the program can be increased or, if this rate is sufficient, there is room for additional computations which may improve the results.

The use of multiple processors becomes a necessity when the minimum set of computations to be performed is given, but the time needed to compute them on one processor is unacceptably long. Then the only way to reduce this execution time is by using several processors. However, the distribution must be done carefully because communication costs are considerable for any computer architecture.

Since the message-passing mechanism of execution is fairly simple, it is possible to determine precisely the time needed to perform all operations. This is a prerequisite for programming real-time applications, since one of the design goals is not to surpass a given time limit for the program's execution time.

For all these reasons, using the data flow approach is very suited for distributing the computations to multiple processors. Basically, the problems to be solved in the distributing process are:

- Generating a data flow graph representation of the algorithms to be produced
- Adjusting the size of the graph components: partitioning the graph into tasks

- Distributing the tasks to the processors, and
- Scheduling the tasks for execution during run-time.

The prime goal targeted during all operations is to minimize the total execution time.

The first three points are addressed in the project described here, while scheduling is left aside for the reason that the Transputer as target hardware already possesses a simple and efficient run-time scheduler.

As soon as computations involving more than a handful of operations are to be analysed and distributed, it becomes impossible to perform all the operations involved manually. As a consequence, automated tools must be developed to aid the process of finding a good mapping of the computations to the processors.

1.2 Book Outline

The introductory issues mentioned above are addressed in Chapter 2, where the background of multicomputers and real-time systems is described in more detail. The aims of the partitioning process are outlined, together with the restrictions found in reality. The SISAL and OCCAM programming languages are introduced which are used for input and output, respectively. A survey of the single processing steps which are performed for automated partitioning and allocation close the chapter.

In Chapter 3 the characteristics of the algorithms considered for the target applications are investigated. All the main classes of algorithms are addressed: various kinds of controllers (PID controllers, linear state feedback controllers, controllers with estimator), numerical integration of linear and nonlinear differential equations, signal processing algorithms (digital filters, signal transforms), and general matrix computations. For all algorithms extensive references to the literature are given.

The common properties of these algorithms are that they can be formulated to a large extent by using matrix expressions and that no recursion is needed. The absence of recursion is important since it allows us to determine in advance the amount of work to be done. This is an important condition for ensuring in advance that the time constraints for real-time applications can be met. Readers already familiar with signal processing and digital control may choose just to browse through Chapter 3 or to skip it entirely.

Chapter 4 addresses the first steps in the automated allocation process: the data flow graph generation, the communication volume analysis, and some modifications of the graphs. Then the communication and computation cost analysis is described. The cost tables apply to the Transputer and are based on a newly developed communication cost model and execution cost measurements. These models apply to any message-passing multicomputer.

In Chapter 5 a method is described for partitioning the graph into tasks. The aims of partitioning are to decrease the number of execution units and to reduce communication costs, while preserving all the parallelism in the data flow graph. An extensive survey is given of approaches for partitioning described in the literature.

The way the tasks are allocated to the processors available in the system is treated in Chapter 6. After briefly addressing the topic of dynamic and static scheduling, the general static task allocation problem is formulated and the special conditions for real-time computations are given. Following an extensive survey of solutions proposed in the literature, the new two-phase linear clustering approach is described.

Basically, this new approach consists of switching the optimization goal at a certain point of processing. In the first phase, the emphasis lies on joining heavily communicating tasks on the same processor and thus eliminating costly interprocessor communication. In the second phase, the optimization tends towards reducing the number of links among the processors and the number of processors utilized until the allocated computations fit the specified hardware. The optimization goal is changed as soon as the average number of tasks reaches the value specified by the user. The influence of this allocation parameter on the final processor configuration is investigated experimentally, as described later in Chapter 7.

One special feature of the allocation procedure is that, *a priori*, no structure of the interconnections among the processors is assumed. The available links are connected according to the specific needs of the application algorithm. Through this optimized resource utilization bottlenecks in the communication are reduced.

Following some general reflections on the structure of a translator, the details of the generation of OCCAM code for the Transputers from the partitioned and allocated data flow graph are described.

Chapter 7 presents case studies of partitioning and allocating the algorithms introduced in Chapter 3. First, all steps of the allocation process are illustrated using a small scalar product as an example. For each algorithm, several examples of different size are processed. The influence of the allocation parameter on the number of processors used and on the estimated total execution time is explored.

The results show that for most examples the shortest execution time is found for allocation parameter values of about one fourth to one eighth of the number of tasks to be allocated. The estimated parallel execution time is lowered to as little as one third of the serial execution time, typically using two to four processors.

The proposed method works well for all but one of the types of algorithms considered. The exception is the fast Fourier transform (FFT). In the FFT, either an inherently sequential execution results or the number of tasks becomes so large (to several thousand) that the processing time is so long that this approach becomes unfeasible. However, this restriction results only from the implementation and not from the method itself, since the FFT is very well suited for parallel implemetation.

Modifications of the task building process are suggested in Chapter 8 as future work together with a review of computer systems other than Transputers.

Appendix A contains the descriptions of the multicomputer and of the pertinent versatile high-speed data acquisition system used for this work. Appendix B lists in detail the processing steps to be performed for partitioning and allocating applications for parallel processing.

CHAPTER 2

Problem Outline

2.1 Types of Multiple-processor Systems

Since the advent of the first multiple-processor computers in the 1960s, a multitude of architectures has been developed for such systems. The first systematic architectural classification by Flynn ([Flynn 66]) is not unique in every respect, but is still widely used. Other, more subtle classification schemes have been proposed by Feng, ([Feng 72]), Händler ([Händle 77]), Giloi ([Giloi 83]), and Skillicorn ([Skilli 88]).

Flynn classifies the computing systems according to the hardware provided to service the instruction and data streams. He gives the following four machine organizations:

- Single instruction stream-single data stream   (SISD)
- Single instruction stream-multiple data stream   (SIMD)
- Multiple instruction stream-single data stream   (MISD)
- Multiple instruction stream-multiple data stream   (MIMD)

A small system such as a workstation is a typical example of the SISD class of computers. In its single processor, the control unit sends one stream of decoded instructions to the processor units for execution. These units receive one stream of data from memory and return the results to memory.

The converse of single instruction and single data streams are multiple instances of the functional units processing data and instructions, as used in the MIMD class of computers. Each component of the computer can be regarded as a complete SISD machine. Depending on the organization of the memory, these computers are further classified as shared-memory and distributed-memory machines. Shared-memory computers are sometimes also called multiprocessor systems, whereas for distributed-memory computers the term multicomputer has been coined ([AthSei 88]), which will be used throughout this book. However, no rigid and universally acknowledged terminology has evolved so far for this area.

In multicomputers, the only way for the single computers or processing elements (PEs) to communicate with each other is by sending messages to each other, since no globally accessible memory space exists. This eliminates access collisions to memory, but can slow down data transfer depending on the organization and bandwidth of the interconnection network. In any case, some kind of protocol is used to handle timely and correct data exchange.

For signal processing systems, the so-called Harvard architecture is widely used (see [Lee 88], [Lee 89]). There, the memory is split into two parts containing the program and the data to be processed, respectively. Each part is accessible over a separate data and address bus. This architecture allows independent and concurrent access to memory, thus speeding up program execution.

The systems considered so far all belong to the von Neumann family of computers in which instructions are executed sequentially as controlled by a program counter. They are also called *control flow* computers. A completely different approach is used in *data flow* computers. The basic concept is to enable the execution of an instruction whenever its operands become available. Thus no program counters are needed in data-driven computations, and instructions in a program are not artificially ordered in a linear sequence.

Theoretically, maximal concurrency can be exploited in data flow machines, constrained only by the hardware resource availability. In practice, however, expensive control mechanisms are needed to check the availability of data for an operation and then to schedule it. Nevertheless, several data flow computers have been built (see [Veen 86], [Gurd 87], [Dennis 80], [YuShYa 90], and [GauBic 91] with an extensive bibliography).

Sequential programs are inappropriate for defining computations on data flow computers since they impose an artificial and partly unnecessary order upon the statements. However, through intelligent analysis sequentially formulated programs are also implementable efficiently on data flow machines (see [BeJoPi 89]). Hence, data flow graphs (see Section 2.6) are used to express the operations to be performed. These graphs contain only the operations, drawn as nodes, and the data dependencies among the operations, shown as arrows or edges. An algorithm expressed by means of a data flow graph is formulated independently of the machine architecture on which it is eventually executed.

Even if no data flow computer is used to execute the operations defined by the data flow graph, formulating the problem in this way offers the great advantage that all parallelism is revealed. With conventional programming languages, obtaining the same amount of parallelism from the program is almost impossible, even when applying the most sophisticated compiling techniques.

Computer systems incorporating multiple processing elements are called inhomogeneous when the PEs are of different type, and homogeneous when all PEs are identical.

Of all the non-data flow architectures, the message-passing multicomputer possesses the most appropriate architecture for executing a data flow graph. The graph's edges can be created by sending messages, and clusters of nodes can be implemented as tasks on the computers. With a simple scheduler running on each computer and appropriate message-handling software, the computations execute in a self-synchronizing manner.

For the project described in this book, a homogeneous message-passing multicomputer consisting of INMOS T800 Transputers ([Whitby 85], [INMOS 89], described in Appendix A) was chosen for the following reasons:

- The data flow graph principle matches well the computational model created by the Transputer
- Message passing is supported in hardware and software
- A fast and efficient run-time scheduler exists for parallel tasks
- High computation power in both fixed-point and floating-point arithmetic
- Complete processor boards are available off the shelf, thus eliminating the need for own hardware development
- Relatively inexpensive hardware

The area of parallel computer architecture is covered in depth by a variety of books and articles, starting with the classical books by Hwang and Briggs ([HwaBri 84]) and Hockney and Jesshope ([HocJes 81]). Other valuable references are [AlmGot 89], [Anders 89], [Duncan 90], and [Trelea 90], together with the surveys of a large number of actual systems given in [DeCega 89] and [TreWil 91].

2.2 Real-time Systems

Real-time information processing is defined as "... the processing of data by a computer in connection with another process outside the computer according to time requirements imposed by the outside process" ([IEEE 88]). No absolute values of the response time of a system are contained in the definition. A system receiving a set of input data every hour and computing results over a period of half an hour could thus be considered a perfect real-time system.

However, in this context fast real-time systems are considered with cyclic (non-terminating) programs involved in industrial control and signal processing. The task turnaround time, i.e., the time elapsed between the arrival of two consecutive sets of input data, is assumed to be in the range of seconds at most, possibly down to the sub-microsecond range.

Control systems of any kind belong to the class of hard real-time systems ([StaRam 88]). They are characterized by the fact that there will be severe consequences if the answer of the system is either incorrect or not in time. Examples of hard real-time systems are process control systems, avionics and flight control systems, or robot and vision systems.

Ensuring that the timing and logical correctness requirements are met is a difficult task. Expensive dynamic pre-emptive scheduling algorithms are necessary for computers with varying load and for general purposes. If the computers are used as dedicated systems, with a known set of tasks to be processed, then scheduling becomes easier. It may then even be possible to use so-called static scheduling, where the task execution order is fixed prior to run time.

While static scheduling saves considerable expenses at run-time, it also needs substantial efforts to establish a good estimate of the task-execution times for determining a good task organization without introducing unnecessary delays. Particularly with multicomputers, the quality of the execution estimates drops rapidly when unforeseen communication delays are considered.

The approach taken in this project is to keep scheduling costs low by using the data flow principle to synchronize the task execution. When analysing the data flow of the application programs in advance, a partial order is determined for the tasks, as opposed to the total order imposed by a fully static schedule. A partial order means that only the precedence relations among the tasks are defined, not the exact times of execution.

The tasks are then distributed to the PEs in accordance with the data dependencies. A simple and efficient round-robin scheduler administers the tasks of each PE at run-time and schedules them as soon as all input values for a task have arrived. The tasks themselves are executed in an conventional sequential way.

Therefore, using the data flow principle for organizing the sequence of the computations has the advantage that only a cheap dynamic scheduler is necessary. Additionally, no exact estimates of the task execution times need be computed off-line, since knowing the tasks' starting times is irrelevant due to the self-synchronization property of the program execution.

For real-time systems, the utilization of multiple-processor computers is thus the obvious choice in order to lower the total computation time of an application program ([RoHaMe 88]). Examples of systems used in control applications are described in [KirKau 84] and [Shaffe 89]. Systems programmed by using the data flow approach are discussed in [Barkho 87], [Lent 87], and [Campbe 85] and [GaVeTu 85] for the Hughes Data Flow Machine (HDFM). Transputers have been applied for real-time control in projects described in [Flemin 88], [Leppäl 87], and [MaIrLi 89].

2.3 Aims of Optimization

When chosing a multicomputer for creating a real-time application, the primary goal is to achieve a minimum execution time, or, in other words, the processing rate of the input signals is maximized for a cyclic execution of the task set.

The computation time is minimized if parallelism is maximized. As a consequence, after analysing the application algorithm, each program section executable in parallel should be assigned a processor of its own. However, the utilization of the processing elements should be maximized and the computational load distributed as evenly as possible among the PEs.

Since the processing elements of the multicomputer communicate by message-passing, the interconnection network should link all PEs directly to each other at a maximum speed, thus minimizing communication delays.

For the development phase, the formulation of the application algorithms to be programmed should be both appropriate and easy. No manual transformation from one representation to another should be necessary in the path between formulating the computations and generating executable code for the target machine. Also, if at all possible, the development cycle time should be short.

2.4 Trade-offs between Aims and Feasibility

The actual achievement of an optimum performance of a multicomputer with n processing elements depends on many details. Achieving linear speed-up is much more difficult for control and signal processing applications than for, say, linear algebra algorithms, which scale for any size of input data. There, parallelism increases with growing problem size, and more processors can be kept busy. With signal processing applications, however, the problem size is fixed, e.g., it is confined to a given number of state variables of a controller, or of signal samples to be transformed.

Therefore, for such a problem an upper bound of parallelism exists which cannot be surpassed. Consequently, the program execution time will decrease to a lower limit when more PEs are employed. When this limit is reached, it does not make sense to try to utilize more PEs since they cannot be kept busy with useful work. Moreover, not only does the processor idle times increase, but the communication overhead may also rise, thus increasing the total program execution time.

Consequently, for a given application problem there is an optimum number of processing elements which yields the minimum execution time. PE utilization is the best possible, and communication among the PEs is minimal. In particular, it will happen that fewer PEs are utilized than are available in the system, since using more does not yield any advantages. The opposite may also happen. When there are not enough PEs for optimum mapping, parallelism must be sacrificed in order to find a solution realizable on the given computer.

The difficult problem of finding this best assignment of the computations to the PEs is dealt with in the subsequent chapters of this book.

It is technically difficult to produce a network which interconnects all PEs at a very high speed. Very high-speed networks are expensive to build. Therefore slower networks incurring substantial delays are used. In most multicomputers, the PEs possess only few fast serial links, which connect them to their neighbours. In hypercube computers, this feature is the essence of the architecture. Communication to distant PEs is routed along a path passing intermediate processing nodes.

This kind of communication network has two major disadvantages: first, all the intermediate PEs are burdened to some extent with additional message routing and passing jobs, and second, the transmission time includes an *a priori* unpredictable component due to the unknown number of hops.

Therefore, in this project the four serial links of the T800 Transputer are used only for communication to the adjacent PEs. A partition of the data flow graph representing the application algorithm has to be devised such that this distance constraint is met. Since the limited number of links does not allow mapping the edges of the data flow graph one-to-one on the links, additional measures such as link multiplexing and link fusion have to be taken, as outlined in Chapter 6. Through this measure, the flexibility of the multicomputer is lowered, and the difficulty of finding a good mapping of the problem on the computer is increased, but the advantage of having predictable communication times is worth sacrificing some flexibility.

The recently announced T9000 Transputer possesses multiplexing hardware to allow any number of processes to use each link, so that links can be used transparently ([INMOS 91]). Tasks are then linked by *virtual channels* which are managed by the virtual channel processor. Communication among the tasks is handled by a packet switching network implemented on the physical links. However, this again introduces unpredictable message delays.

2.5 The SISAL and OCCAM Programming Languages

Defining the application algorithm to be implemented on the parallel machine needs a suitable tool. While a graph editor for drawing the algorithms as a signal-flow diagram creates easily understood pictures, the editing process itself very soon becomes tedious and error-prone. Moreover, the amount of parallelism displayed in such a diagram is limited by the user's perception of the computations and does not exhibit the true relations.

Therefore, a tool was chosen which allows the formulation of the computations in an easily understandable programming language, SISAL. Subsequently, the program code is converted to a data flow graph representation. The fact that SISAL is a high-level language similar to well-known computer languages such as Pascal eases familiarization considerably.

SISAL (Streams and Iterations in a Single Assignment Language [MGSkAl 85]) is a general-purpose applicative language intended for use on both conventional and new multiprocessor systems ([FeoCan 90]). It follows the Single Assignment axiom, which says that each variable must be assigned a value only once. (Well-established programming languages like Pascal or Fortran allow the same variable to be used for completely independent purposes at several locations in the program, thus creating artificial data dependencies.) Through the Single Assignment principle, the existing data dependencies are identified and shown in the data flow graph.

The SISAL language grew out of work done for VAL (Value-oriented Algorithmic Language [AckDen 79]) which is described together with other data flow languages and related topics in [Ackerm 79].

However, some restrictions are imposed on the use of the full SISAL language and these are shown in Table 2.1. The reason for the restrictions imposed on the data types is that only "well-behaved" application algorithms are considered which do not generate or require values indicating an error. The compound data types Stream and Union are difficult to implement in OCCAM on the Transputer. The type Record can be emulated with some manipulations, but it is rarely used in signal processing applications.

```
function controlalgorithm(Inputs: array[type];
    NumberOfInputs: integer;
    StatesIn: array[type]; NumberOfStates: integer;
returns array[type]; array[type])

...   % function code

end function
```

Figure 2.1 Header of Function control algorithm

Table 2.1 *Restrictions on the Use of SISAL*

No use of the data type: Error
No use of the compound data types: Stream, Union, Record
No recursive function calls

Since the Transputer assigns static workspace only to its processes, recursive function calls are not possible. Therefore, they are barred from use in SISAL.

The function containing the computations to be parallelized has the mandatory name *controlalgorithm*. It is assumed that at each sampling instance one set of input values is assembled and passed to the function *controlalgorithm*. Its header has been defined to appear as shown in Figure 2.1.

The parameters have the following meaning:

Inputs	array of input values
NumberOfInputs	number of elements in the input array
StatesIn	array of internal states returned in last call
NumberOfStates	number of elements in the state array

The first array returned contains the output values computed by the function and the second array represents the state information which is passed to the next invocation of *controlalgorithm*.

After the computations have been analysed and distributed to the PEs, they are converted to the OCCAM programming language for compilation on the Transputers. OCCAM ([INMOS 88]) "is a high level language, designed to express concurrent algorithms and their implementation on a network of processing components". In OCCAM, *processes* are defined which execute concurrently and communicate with other processes through *channels*. This allows structuring the application clearly, and exhibiting the inherent parallelism.

Concurrency and communication are the prime concepts of the OCCAM model which is based on EPL (Experimental Programming Language) by May, Taylor, and Whitby-

12 *Problem outline*

Strevens ([MaTaWh 78]) and on CSP (Communicating Sequential Processes) by Hoare ([Hoare 78]). This model captures the hierarchical structure of a system by allowing an interconnected set of processes to be regarded as a unified, single process. It is therefore very well suited to represent program structures derived from data flow graphs.

The most recent proposal for a revision of OCCAM ([Barret 90]), called OCCAM91, introduces a more comprehensive type system and support for a modular programming style. Additionally, the facility for sharing objects between processes is provided. These features will be supported efficiently by the new T9000 Transputer.

2.6 Data Flow Graphs

The principle of data flow graphs is to let the availability of data determine the order of execution of the statements. As soon as an operation has received its input values, it is ready to execute and to generate its result values, which flow to the next, connected statements.

A data flow graph can be described in the same way as an ordinary directed graph. It consists of a set of *nodes* which represents the operations. Only a limited number of well-defined node types exist. The nodes are connected by the set of directed *edges* according to the data dependencies. Edges carry data tokens which represent data of any type.

One of the first data flow models for computations was developed by Adams ([Adams 68]) who used the term "data flow" for the first time. A very influential model was presented by Dennis, Fosseen, and Linderman in 1973 ([DeFoLi 73]). In this

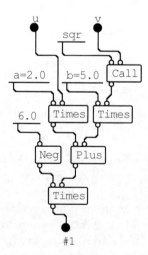

Figure 2.2 Example of a Simple Data Flow Graph in IF1

model, each edge could carry only one token at a given moment in time. To model asynchronous computations, tagged tokens were introduced by Arvind, Kathail, and Pingali ([ArKaPi 80]). They defined each token to carry an additional tag identifying the invocation of the data flow graph to which it belongs.

For signal processing applications which are computed without using pipelining it is not necessary to introduce tagged tokens. Every execution of the graph must have finished before the next begins.

The data flow graph model used in this project was developed for use as an intermediate format in the compiler for SISAL, hence its name IF1 (Intermediate Form 1, [SkeGla 85]).

Figure 2.2 shows as a simple example the graph representing the expression $y = -6.0 \times (a \times u + b \times v^2)$ with $a = 2.0$, $b = 5.0$.

Data flow graphs expressed in IF1 are hierarchical, since more complex structures such as loops are visible on the top layer as compound nodes. These compound nodes contain several subgraphs, e.g., for the initialization phase, the loop body and the result-gathering section.

2.7 Survey of the Processing Steps for Partitioning and Allocating an Application Program

The steps which are taken to transform an application algorithm from its textual representation to a data flow graph, then to partition it and to map it onto a multicomputer are outlined in the following section. A preliminary description of this process has already been given in [Thoeni 91]. Figure 2.3 shows the processing phases which are described in the following chapters.

From the algorithm expressed as a SISAL program a data flow graph is generated with the standard SISAL compiler. This graph is then analysed for the amount of communication transmitted over its edges. Subsequently, the graph is expanded, which mainly applies to the loop nodes which are unrolled. At the end of the preparatory phase each node is assigned its execution costs which are read from a processor-specific table.

During the partitioning phase the graph's nodes are clustered into tasks. These tasks are carefully formed so that no parallelism in the data flow graph is destroyed.

In the allocation stage the tasks are distributed to the real computer system with a limited number of processing elements and links among them. The allocation is carried out in two phases. In the first phase, the emphasis lies more on reducing interprocessor communication by clustering heavily communicating tasks on the same PE. The second phase attempts to find an optimum mapping which yields the lowest execution time, given the constraints imposed by the real multicomputer system. As a result, a list of the connections among the PEs is generated and the number of PEs used is determined.

For the allocation phase a parameter is requested from the user. Through this parameter the moment of switching from pure minimization of communication (stage one) to respecting the hardware constraints (stage two) is influenced.

14 *Problem outline*

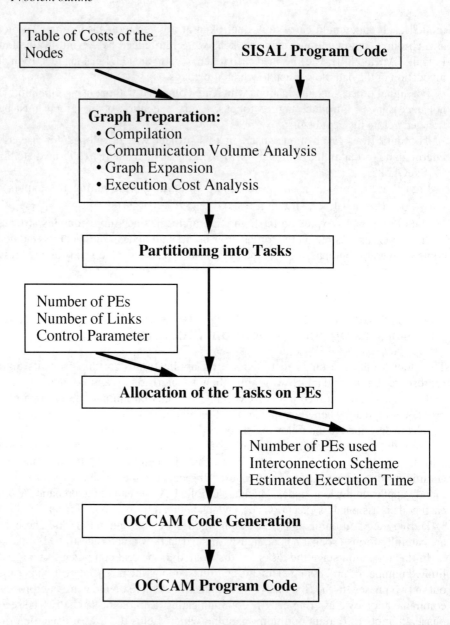

Figure 2.3 Processing Steps for Partitioning and Allocating

The partitioned and allocated tasks of the data flow graph are then translated into OCCAM code in a straightforward manner. The tasks are modelled as OCCAM processes. All tasks located on one Transputer are defined as executable in parallel. The edges connecting tasks on different PEs are mapped to the links connecting the Transputers. This OCCAM program is then passed through the standard OCCAM compiler.

The numerical complexity of this partitioning and mapping process is relatively high, such that the computation time spent on preparing the executable program may be considerable. However, the effort is justified in that the program execution time at runtime is lowered to the minimum by solving the very complex partitioning and mapping problem.

2.8 References

[Ackerm 79]   W.B. Ackerman, "Data Flow Languages," in *Proc. National Computer Conference*, vol. 48, pp. 1087-1095, 1979.

[AckDen 79]   W.B. Ackerman and J.B. Dennis, "VAL–A Value-Oriented Algorithmic Language: Preliminary Reference Manual," MIT Laboratory for Computer Science, Technical Report MIT/LCS/TR-218, June 1979.

[Adams 68]   D.A. Adams, "A Computation Model with Data Flow Sequencing," Computer Science Dept., Stanford University, Technical Report No. CS 117, December 1968.

[AlmGot 89]   G.S. Almasi and A. Gottlieb, *Highly Parallel Computing*, Redwood City, CA a.o.: The Benjamin/Cummings Publishing Co., 1989.

[Anders 89]   A.J. Anderson, *Multiple Processing: A Systems Overview*, Englewood Cliffs, NJ: Prentice Hall, Inc., 1989.

[ArKaPi 80]   Arvind, V. Kathail and K. Pingali, "A Dataflow Architecture with Tagged Tokens," MIT Laboratory for Computer Science, Technical Memorandum No. 174, 1980.

[AthSei 88]   W.C. Athas and C.L. Seitz, "Multicomputers: Message-Passing Concurrent Computers," *Computer*, vol. 21, no. 8, pp. 9-24, 1988.

[Barkho 87]   S. Barkhordarian, "RAMPS: A Realtime Structured Small-Scale Dataflow System for Parallel Processing," in *Proc. Int. Conf. Parallel Proc.*, 1987, pp. 610-613.

[Barret 90]   G. Barrett, "The Development of OCCAM: Types, Classes and Sharing," in: OUG-13, Real-Time Systems With Transputers, H.S.M. Zedan (ed.), Amsterdam a.o.: IOS Press, 1990, pp.119-147.

[BeJoPi 89]   M. Beck, R. Johnson and K. Pingali, "From Control Flow to Dataflow," Department of Computer Science, Cornell University, Ithaca, NY, Technical Report TR 89-1050, October 1989.

[Campbe 85]   M.L. Campbell, "Static Allocation for a Data Flow Multiprocessor," in *Proc. Int. Conf. Parallel Proc.*, 1985, pp. 511-517.

[DeCega 89]   A.L. DeCegama, *The Technology of Parallel Processing, Volume 1: Parallel Processing Architectures and VLSI Hardware*, Englewood Cliffs, NJ: Prentice Hall, Inc., 1989.

[Dennis 80]   J.B. Dennis, "Data Flow Supercomputers," *Computer*, vol. 13, no. 11, pp. 48-56, 1980.

[DeFoLi 73]   J.B. Dennis, J. Fosseen and J. Linderman, "Data Flow Schemas," in: *Proc. International Symposium on the Theory of Programming*, pp. 187-216, 1973. New York, a.o.: Springer-Verlag, Lecture Notes in Computer Science, vol. 5, 1974.

[Duncan 90]   R. Duncan, "A Survey of Parallel Computer Architectures," *Computer*, vol. 23, no. 2, pp. 5-16, 1990.

[Feng 72]   T.Y. Feng, "Some Characteristics of Associative Parallel Processing," in *Proc. Sagamore Comp. Conference*, 1972, pp. 5-16.

[FeoCan 90] J.T. Feo and D.C. Cann, "A Report on the Sisal Language Project," *J. Parallel and Distr. Computing*, vol. 10, no. 4, pp. 349-366, 1990.

[Flemin 88] P.J. Fleming, *Parallel Processing in Control: The Transputer and Other Architectures*, IEE Control Engineering Series, vol. 38. London: Peter Peregrinus Ltd., 1988.

[Flynn 66] M.J. Flynn, "Very High-Speed Computing Systems," *Proc. IEEE*, vol. 54, pp. 1901-1909, 1966.

[GauBic 91] J.-L. Gaudiot and L. Bic, *Advanced Topics in Data-Flow Computing*, Englewood Cliffs, NJ: Prentice Hall, Inc., 1991.

[GaVeTu 85] J.-L. Gaudiot, R.W. Vedder, G.K. Tucker, et al., "A Distributed VLSI Architecture for Efficient Signal and Data Processing," *IEEE Trans. Computers*, vol. 34, no. 12, pp. 1072-1087, 1985.

[Giloi 83] W.K. Giloi, "Towards a Taxonomy of Computer Architecture Based on the Machine Data Type View," in *Proc. IEEE Conf. Parallel Proc. and Computer Arch.*, pp. 6-13, 1983.

[Gurd 87] J.R. Gurd, "Dataflow Architectures," in: *Major Advances in Parallel Processing*, Chris Jesshope (ed.), Aldershot: Gower Technical Press Ltd., 1987.

[Händle 77] W. Händler, "The Impact of Classification Schemes on Computer Architecture," in *Proc. Int. Conf. Parallel Proc.*, 1977, pp. 7-15.

[Hoare 78] C.A.R. Hoare, "Communicating Sequential Processes," *Communications of the ACM*, vol. 21, no. 8, pp. 666-677, 1978.

[HocJes 81] R.W. Hockney and C.R. Jesshope, *Parallel Computers*, Bristol: Adam Hilger Ltd., 1981.

[HwaBri 84] K. Hwang and F.A. Briggs, *Computer Architecture and Parallel Processing*, New York, a.o.: McGraw-Hill Book Company, 1984.

[IEEE 88] Institute of Electrical and Electronics Engineers, Inc., *Standard Dictionary of Electrical and Electronics Terms*, Fourth Edition. ANSI/IEEE Std. 100-1988. New York 1988.

[INMOS 88] INMOS Ltd., $occam^{®}$ *2 Reference Manual*, New York, a.o.: Prentice Hall, Inc., 1988.

[INMOS 89] INMOS Ltd., *The Transputer Databook*, Second Edition 1989.

[INMOS 91] INMOS Ltd., *The T9000 Transputer Products Overview Manual*, 1991.

[KirKau 84] H.D. Kirrmann and F. Kaufmann, "Poolp–A Pool of Processors for Process Control Applications," *IEEE Trans. Computers*, vol. 33, no. 10, pp. 869-878, 1984.

[Lee 88] E.A. Lee, "Programmable DSP Architectures: Part I," *IEEE ASSP Magazine*, vol. 5, no. 4, pp. 4-19, 1988.

[Lee 89] E.A. Lee, "Programmable DSP Architectures: Part II," *IEEE ASSP Magazine*, vol. 6, no. 1, pp. 4-14, 1989.

[Lent 87] B. Lent, "Data Flow Driven Computer for Embedded Control Systems," *Microprocessors & Microprogramming*, vol. 19, pp. 385-399, 1987.

[Leppäl 87] K. Leppälä, "Utilization of Parallelism in Transputer-Based Real-Time Control Systems," *Microprocessors & Microprogramming*, vol. 21, pp. 629-636, 1987.

[MaIrLi 89] L.P. Maguire, G.W. Irwin and G. Lightbody, "Mapping Control Algorithms onto Transputer Arrays," Colloquium on Transputer Applications, organized by the Professional Group C2, 13 November 1989. IEE Digest No. 1989/129.

[MaTaWh 78] M.D. May, R.J.B. Taylor and C. Whitby-Strevens, "EPL–An Experimental Language for Distributed Computing," in *Proc. 1978 Conf. Distributed Processing: Trends and Applications*, Gaithersburg, MD, pp. 69-71, May 1978.

[MGSkAl 85] J. McGraw, S. Skedzielewski, S. Allan, et al., "SISAL: Streams and Iteration in a Single Assignment Language, Reference Manual," Version 1.2, Lawrence Livermore National Laboratory Report LLL/M-146 Rev. 1, 1 March 1985.

[RoHaMe 88] J.B.G. Roberts, J.G. Harp, B.C. Merrifield, et al., "Evaluating Parallel Processors for Real-Time Applications," *Parallel Computing*, vol. 8, pp. 245-254, 1988.

References

[Shaffe 89] P.L. Shaffer, "Experience with Implementation of a Turbojet Engine Control Program on a Multiprocessor," in *Proc. American Control Conf.*, 1989, pp. 2715-2720.

[SkeGla 85] S. Skedzielewski and J. Glauert, "IF1, An Intermediate Form for Applicative Languages," Version 1.0, Lawrence Livermore National Laboratory Report M-170, July 31, 1985.

[Skilli 88] D.B. Skillicorn, "A Taxonomy for Computer Architectures," *Computer*, vol. 21, no. 11, pp. 46-57, 1988.

[Srini 86] V.P. Srini, "An Architectural Comparison of Dataflow Systems," *Computer*, vol. 19, no. 3, pp. 68-88, 1986.

[StaRam 88] J.A. Stankovic and K. Ramamritham (eds.), *Hard Real-Time Systems (Tutorial)*, Washington D.C.: IEEE Computer Society Press, 1988.

[Thoeni 91] U.A. Thoeni, "RTPSP: A Real-Time Parallel Signal Processing Environment for Fast Homogeneous Message-Passing Multicomputers," in *Proc. Int. Conf. Parallel Proc.*, pp. II-150–II-157, August 1991.

[Trelea 88] P.C. Treleaven, "Parallel Architecture Overview," *Parallel Computing*, vol. 8, pp. 59-70, 1988.

[TreWil 91] A. Trew and G. Wilson (eds.), *Past, Present, Parallel: Computing Systems, A Survey of Available Parallel Computing Systems*, London, Berlin, Heidelberg a.o: Springer-Verlag, 1991.

[Veen 86] A.H. Veen, "Dataflow Machine Architecture," *ACM Computing Surveys*, vol. 18, no. 4, pp. 365-396, 1986.

[Whitby 85] C. Whitby-Strevens, "The Transputer," in *Proc. 12th Ann. Symposium Comp. Architecture*, 1985, pp. 292-300.

[YuShYa 90] T. Yuba, T. Shimada, Y. Yamaguchi, *et al.*, "Dataflow Computer Development in Japan," in: *Proc. 1990 Int. Conf. Supercomputing*, Amsterdam, 11-15 June 1990, pp. 140-147. New York: ACM Press, Inc., 1990.

CHAPTER 3

Survey: Algorithms for Control and Signal Processing and their Characteristics

3.1 Introduction

Mathematical problems can be formulated in a variety of ways, e.g., with a formula, verbally, or even graphically. For solving them, however, an algorithm has to be formulated, and is a set of rules that is supposed to terminate in a finite number of steps on a computer. Generally, more than one algorithm exists to solve a specific problem. Depending on its formulation, each can be evaluated numerically in different ways. As computer arithmetic is of finite accuracy only, different results can evolve, depending on the algorithm used and its evaluation. The choice of the best algorithm for a given problem and for a specific computer is a difficult task and depends on many details.

In this chapter, some design issues of algorithms used in control applications and in signal processing are discussed and the general properties of these algorithms are outlined. The following algorithms will be reviewed:

1. Controllers   single-input single-output (SISO) controller
                               digital PID controller
                               general dynamic controller
                     state space controller
                               linear state feedback
                               linear state feedback with estimator
                               optimal estimator: the Kalman filter
2. General systems   linear system
                     nonlinear system
3. Numerical integration for solving nonlinear system equations
4. Signal processing   filtering
                            FIR filters
                            IIR filters
                   signal transforms
                            discrete Fourier transform (DFT)

5. Matrix computations
  fast Fourier transform (FFT)
  convolution, correlation, spectrum estimation
  vector-vector operations
  matrix-vector operations
  matrix-matrix operations
  matrix inversion
  solution of linear equation systems

Many of these algorithms can be expressed by matrix computations. They are called regular iterative algorithms (RIA) ([RaoKai 87], [Rao 85]) due to their very regular structure. One property of the controllers and filters treated here is that they are causal systems, i.e., the output values depend only on past input and state values.

In this context, the focus is on the structure of the algorithms, i.e., how the input and the output values of one control step are dependent on each other. Numerical and other implementational aspects are of secondary importance for the investigation of the potential parallelism in those algorithms. For such issues see [Goldbe 91], [Hansel 87], [Allen 85], or [Morone 83].

3.2 Notation for the Description of Systems

The following notation for describing dynamic, continuous or discrete systems (plants or controllers) will be used throughout this book. It follows the terminology used in [FraPow 90], with some extensions.

3.2.1 Linear Systems

State space representation:

continuous:
$$\dot{x}(t) = F\,x(t) + G\,u(t) \tag{3.1}$$
$$y(t) = H\,x(t) + J\,u(t) \tag{3.2}$$

discrete:
$$x_{k+1} = \Phi\,x_k + \Gamma\,u_k \tag{3.3}$$
$$y_k = H_d\,x_k + J_d\,u_k \tag{3.4}$$

Transfer function:

continuous:
$$G(s) = \frac{b_q\,s^q + b_{q-1}\,s^{q-1} + \ldots + b_1\,s + b_0}{s^n + a_{n-1}\,s^{n-1} + \ldots + a_1\,s + a_0} \tag{3.5}$$

discrete:
$$G(z) = \frac{b_q z^q + b_{q-1} z^{q-1} + \ldots + b_1 z + b_0}{z^n + a_{n-1} z^{n-1} + \ldots + a_1 z^1 + a_0} \quad (3.6)$$

The coefficients a_i and b_i, respectively, of the transfer functions appear in the differential equation for continuous systems and also in the difference equation describing a discrete system.

Discrete plant with stochastic disturbances:

plant $\quad\quad\quad\quad x_{k+1} = \Phi x_k + \Gamma u_k + \Gamma_1 w_k \quad (3.7)$

measurements $\quad y_k = H_d x_k + v_k \quad (3.8)$

The process noise w_k and the measurement noise v_k are uncorrelated random sequences with zero mean and known covariances.

3.2.2 Nonlinear Systems

Nonlinear systems are described in general terms of the state x and of the input u. However, only time-invariant systems will be treated so that the system equation is not an explicit function of the time t.

Continuous nonlinear plant $\quad \dot{x}(t) = f(x(t), u(t)), \quad x(0) = x_0 \quad (3.9)$

$$y(t) = g(x(t), u(t)) \quad (3.10)$$

The discrete equivalent of the continuous nonlinear plant will not be used. Instead, a linearized continuous system will be transformed to a discrete equivalent or the nonlinear equations will be integrated directly (see Section 3.5).

A nonlinear system defined by 3.9 and 3.10 can be linearized along a nominal state and input x_0 and u_0, respectively. The resulting system matrices consist of the partial derivatives (the Jacobian matrices).

3.3 Linear Controllers

3.3.1 Single-input Single-output (SISO) Controllers

A system is called single-input single-output if the dimension of its input and output vector is one. This means m = 1 and p = 1 for the respective vectors y and u. However, the order of the system, that is, the dimension n of the state vector x, is not related to the dimensions m and p and may be any number.

The PID Controller

A well-known single-input single-output controller is the PID (proportional plus integral plus derivative) controller with the continuous transfer function

$$G(s) = K_P + \frac{K_I}{s} + K_D s = \frac{K_D s^2 + K_P s + K_I}{s} \tag{3.11}$$

When discretized using the bilinear transform ([Tustin 47])

$$s = \frac{2}{T} \frac{z-1}{z+1}$$

the discrete transfer function becomes

$$G(z) = \frac{U(z)}{E(z)} = \frac{b_2 z^2 + b_1 z + b_0}{z^2 + a_1 z^1 + a_0} \tag{3.12}$$

with the corresponding coefficients

$$a_0 = -1, \quad a_1 = 0, \qquad b_0 = -K_P + \frac{K_I T}{2} + \frac{2 K_D}{T}$$

$$b_1 = K_I T - \frac{4 K_D}{T}, \qquad b_2 = K_P + \frac{K_I T}{2} + \frac{2 K_D}{T}$$

When computing the output of the controller 3.11 in a sampled control loop, the next value u_k of the control signal is expressed by the difference equation

$$u_k = b_2 e_k + b_1 e_{k-1} + b_0 e_{k-2} - a_1 u_{k-1} - a_0 u_{k-2} \tag{3.13}$$

Reformulated as a scalar product, equation 3.13 reads

$$u_k = [b_2 \; b_1 \; b_0] \begin{bmatrix} e_k \\ e_{k-1} \\ e_{k-2} \end{bmatrix} - [a_1 \; a_0] \begin{bmatrix} u_{k-1} \\ u_{k-2} \end{bmatrix} = b^T y_{02} - a^T u_{12} \tag{3.14}$$

with the vectors

$$a = [a_1 \; a_0]^T, \qquad b = [b_2 \; b_1 \; b_0]^T$$

describing the controller. The vectors

$$e_{02} = [e_k \; e_{k-1} \; e_{k-2}]^T, \quad u_{12} = [u_{k-1} \; u_{k-2}]^T$$

denote the past values of the error e and of the control vector u which are required for the computation of the next value of u_k.

General Dynamic SISO Controller

Depending on the type of the plant, dynamic systems other than the PID controller can be used as compensator. Such a general dynamic system of order n with input e and output u is described by the transfer function 3.6. For the system to be causal it is required that $q \leq n$.

The next control output u_k is computed by

$$u_k = b^T e_{0q} - a^T u_{1n} = \sum_{i=0}^{q} b_{q-i} e_{k-i} - \sum_{i=1}^{n} a_{n-i} u_{k-i} \qquad (3.15)$$

with the corresponding vectors

$$a = [a_{n-1} \ldots a_0]^T, \qquad b = [b_q \ldots b_0]^T$$

$$e_{0q} = [e_k \ldots e_{k-q}]^T, \qquad u_{1n} = [u_{k-1} \ldots u_{k-n}]^T$$

Again, u_k is the difference of two scalar products.

3.3.2 Multiple-input Multiple-output (MIMO) Controllers

In order to achieve better dynamic properties of the control loop than with output feedback control, state space controllers are applied. Such controllers are more robust and guarantee better stability of the loop. For the design of such a controller it is necessary to have a description of the plant in the state space. If the whole state vector x_k is available, such a control system consists of a linear state feedback. In general, however, not all the internal states of the plant can be measured. In that case, an observer is needed to supply the necessary information based on an approximate model of the plant.

Linear State Feedback

Assuming that the plant is in the state space form 3.3 and that its entire state vector x_k is available, a linear state feedback can be set up in the form

$$u_k = - K x_k \qquad (3.16)$$

The m × n controller matrix K can be established by pole placement, where the poles λ_i (the solutions of $\det(\lambda I - [\Phi - \Gamma K]) = 0$) of the closed loop are chosen by an appropriate method. Such a design method could rely on some performance measure of the control system, e.g., on the requirements of the settling time or of the step response. Other possibilities for the determination of K are the optimal controller, for example according to the LQG (Linear Quadratic Gaussian) assumption (see [KwaSiv 72], [ÅstWit 90], [FraPow 90]) or a robust version thereof, the LQG/LTR design ([Geerin 86], [Geerin 87]).

The next control input u_k is a linear combination of the state variables determined by a multiplication of the state vector x_k by the controller gain matrix K.

Linear State Feedback with Estimator

Again, the control law is $u_k = -K x_k$. However, if the state vector x_k of the plant is only partially accessible, a mathematical model of the physical plant is used to estimate it. Since the difference between the estimated state $\hat{x}_k$ and the real state x_k should be minimal even with some parameter uncertainties in the model, it is not wise to use the system model Φ, Γ alone for the prediction. Instead, the additional information available through the plant output y_k is utilized to correct the predicted state. This leads to the following equation of the prediction estimator

$$\hat{x}_{k+1} = \Phi \hat{x}_k + \Gamma u_k + L [y_{k+1} - H \hat{x}_k] \qquad (3.17)$$

With this predicted state, the following control input is computed:

$$u_k = -K \hat{x}_k \qquad (3.18)$$

The dynamics of the error $x_{k+1} - \hat{x}_{k+1}$ depend on the matrix $[\Phi - LH]$. Hence they can be influenced in an appropriate way by choosing L accordingly.

Instead of the single matrix times vector product in 3.16, the linear state feedback using an estimator requires further computations of the same type, but again the sum and the difference of vectors appear, as shown in equations 3.17 and 3.18.

Optimal Estimator: The Kalman Filter

Given a plant with stochastic disturbances as noted in equations 3.7 and 3.8, it is possible to determine a time-varying gain L_k in order to obtain an estimated state vector $\hat{x}_{k+1}$. It is optimal in the sense that the variance of the error $x_{k+1} - \hat{x}_{k+1}$ is minimal. This estimator is called a Kalman filter after R.E. Kalman [Kalman 60]. The required relations are (from [FraPow 90], chapter 9) between the measurements

$$\bar{x}_{k+1} = \Phi \hat{x}_k + \Gamma u_k \qquad (3.19)$$

$$M_{k+1} = \Phi P_k \Phi^T + \Gamma_1 R_w \Gamma_1^T \qquad (3.20)$$

$$L_{k+1} = M_{k+1} H_d^T (H_d M_{k+1} H_d^T + R_v)^{-1} \qquad (3.21)$$

and at the update time

$$\hat{x}_{k+1} = \bar{x}_{k+1} + L_{k+1}(y_{k+1} - H\bar{x}_{k+1}) \qquad (3.22)$$

$$P_{k+1} = M_{k+1} - L_{k+1} H_d M_{k+1} \qquad (3.23)$$

This procedure requires the update of the matrices M, L, and P between the sampling times. Whereas for M and P these operations consist mostly of vector additions, vector subtractions and matrix multiplications, the computation of the new gain matrix L incorporates a matrix inversion. A detailed analysis of the numerical effort needed is given in [Mendel 71].

3.4 General Systems

3.4.1 Linear Systems

So far, the controllers described have been either static systems, as the state space controller 3.17, or dynamic systems, as described by a transfer function of the type 3.6 and the observers given in 3.17 and 3.19-3.23. However, dynamic systems described by a transfer function can also be expressed by an equivalent state space notation.

The state x_{k+1} of a linear system described in state space notation can easily be computed by matrix times vector products and vector additions, as shown in Section 3.3. It may be expressed as in equation 3.11:

$$x_{k+1} = \Phi x_k + \Gamma u_k$$

The output of a linear system described by a transfer function is computed for a given input sequence according to the corresponding difference equation 3.15. As shown, it can also be formulated as the difference of two scalar products.

The difference equation for the output u depending on the input e reads as given in 3.15:

$$u_k = b^T e_{0q} - a^T u_{1n} = \sum_{i=0}^{q} b_{q-i} e_{k-i} - \sum_{i=1}^{n} a_{n-i} u_{k-i}$$

3.4.2 Nonlinear Systems

When the states of a real plant must be estimated either for a simulation or for a state feedback this is done with the help of a mathematical model. Real plants can seldom be described by a completely linear model. Slight nonlinearities are often neglected and (hopefully) compensated by the robustness of the controller.

Strong nonlinearities arise from real sensors and actuators such as the lambda probe used in cars to measure the oxygen concentration in the exhaust, or valves in hydraulic applications. Additionally, the actuator signals must be limited primarily for physical reasons, for example in robotics where the actuator motors supply only limited torque.

One approach in the simulation of such systems is to linearize the model in the set point. This is inappropriate when the set point varies greatly. Another approach is the use of several linearized models according to the actual state. However, a smooth transfer from one model to another is critical. Then the full nonlinear model 3.9, 3.10 has to be computed in real time, requiring sufficient computational power.

3.5 Solving Nonlinear System Equations by Numerical Integration

Numerical integration is a demanding subject, since the choice of an appropriate method depends very much on the problem to be solved. However, in connection with control and signal processing, only a limited number of types of problems appear. Additionally, since the computations have to be performed in real-time, not all approaches are feasible for integrating the equations.

Ordinary differential equations are sufficiently exact to model conventional plants such as mechanical, hydraulic, and electrical systems. The equations may be linear or nonlinear, e.g., in robotics they are nonlinear since many trigonometric functions are used. It is always possible to transform the equations to a system of differential equations of first order as given in equations 3.9 and 3.10. The problem to be solved is an initial value problem, i.e., the state x_k at time kT given an initial state x_0 must be computed. Results are required after each sampling interval T, but the computation of intermediate values is allowed if needed.

Standard program libraries provide a variety of codes for solving differential equations numerically. Some of the best-known algorithms are the Runge-Kutta-Fehlberg codes RKF45 ([ShaWat 77]) and DVERK ([HuEnJa 76]). Linear multistep methods are also available through codes based on Adams formulas, for example DVDQ ([Krogh 69]), DIFSUB ([Gear 71]), GEAR ([Hindma 74]), VOAS ([Sedgwi 73]), EPISODE ([ByrHin 75]), STEP ([ShaGor 75]), and LSODE ([Hindma 80]).

As the computations of the state vector x have to be executed in real-time, an upper limit of the computation time exists. This does not necessarily put a restriction on the complexity of the integration method as this can be overcome by exploiting parallelism, but it does eliminate methods for which no upper bound for the number of operations exists. Therefore, no implicit integration methods can be used. The amount of computational work is also undeterminable when explicit methods with variable step size are used. Furthermore, while these methods are popular and very effective for off-line simulation, they are of limited use for real-time estimators because the computed states of the plant's model are needed at the well-defined times kT.

Since only methods with fixed step size can be used, the accuracy of the computations is reduced if the step size is too large, or the computational load becomes unnecessarily large for too small a step size. Above all, it will be difficult to treat stiff problems because stability of the integration becomes a major factor. (A problem is said

Solving nonlinear system equations

to be stiff if the eigenvalues of its Jacobian lie several magnitudes apart (see [ShaGea 79], [Shampi 83]).) In such a case, either the model must be reduced to the minimal order necessary or an appropriate integration method must be found. Methods for general ordinary differential equations are treated in [GuSaTi 85], for nonstiff problems in [HaNøWa 87], and for stiff problems in [HaiWan 91].

Two examples of general integration routines of order 4 (global discretization error $O(h^4)$ with h representing the step size) for nonstiff problems are given in the following sections. Methods with a lower order than 4 should not be used for nonlinear systems due to their considerable integration error. The ordinary Euler integration is of order 1 and the trapezoid and Heun integration (the predictor-corrector method) (see [Schwar 86]) are of order 2. A Runge-Kutta integration method is given as representative of a one-step method, and the Adams-Bashforth algorithm is shown for a multistep method.

The definition of the nonlinear system is

$$\dot{x}(t) = f(x(t), u(t)), \qquad y(t) = g(x(t), u(t)),$$

$$x(0) = x_0$$

(see equations 3.9 and 3.10). Since the aim is to compute x_{k+1} from a given state x_k and the control vector u(t) is constant during the interval $kT \le t < (k+1)T$, the equations given above reduce to the form in 3.24. For clarity, an additional parameter kT for time is introduced in the function f:

$$\dot{x}_k = f(kT, x_k, u_k), \qquad y_k = g(x_k, u_k) \tag{3.24}$$

3.5.1 One-step Method: Fourth-order Runge-Kutta Algorithm

With an n-th-order Runge-Kutta algorithm, the nonlinear continuous ordinary differential equations as given above are solved for time instances t = kT and fixed spacing of the points by T according to the following rule:

$$x_{k+1} = x_k + T \sum_{i=1}^{n} b_i\, f(kT + c_iT,\, X_i,\, u_k) \tag{3.25a}$$

$$X_i = x_k + T \sum_{j=1}^{n} a_{ij}\, f(kT + c_jT,\, X_j,\, u_k) \tag{3.25b}$$

$$i = 1, 2, .. n$$

For explicit methods, as used here, the relation $a_{ij} = 0$ holds for $j \geq i$. The other parameters a_{ij} for a fourth-order method (n=4) are given in Tables 3.1 and 3.2.

Table 3.1 *Coefficients a_{ij} for the Fourth-order Runge-Kutta Integration*

i \ j->	1	2	3	4
1	0	0	0	0
2	$\frac{1}{3}$	0	0	0
3	$-\frac{1}{3}$	1	0	0
4	1	-1	1	0

Table 3.2 *Coefficients b_k, c_k for the Fourth-order Runge-Kutta Integration*

k ->	1	2	3	4
b_k	$\frac{1}{8}$	$\frac{3}{8}$	$\frac{3}{8}$	$\frac{1}{8}$
c_k	0	$\frac{1}{3}$	$\frac{2}{3}$	1

For the fourth-order method, for each time step T four evaluations of the nonlinear function vector f are necessary plus one of the function vector g if the output vector y has to be computed as well. The intermediate result vectors X_j have to be summed during the computation. Thus, computing the next state x_{k+1} is a relatively expensive task which is performed with function evaluations and vector additions.

3.5.2 Multistep Method: Fourth-order Adams-Bashforth Algorithm

If the multiple evaluation of the function f is too expensive, considerable savings can be achieved by using a multistep method. In such a method, several states preceding the last are used to compute the next. This means that just one function evaluation per time step T is necessary and that previously computed results are re-used.

Generally, a multistep method of order n has the form

$$x_{k+1} = x_k + T \sum_{i=0}^{n} \beta_i \, f((k+1-i)T, x_{k+1-i}, u_{k+1-i}) \qquad (3.26)$$

Applied to the order 4 Adams-Bashforth integration method, the values shown in Table 3.3 result for the weighting factors β_i.

Table 3.3 *Coefficients for the Fourth-order Adams-Bashforth Integration*

i ->	0	1	2	3	4
β_i	0	$\frac{55}{24}$	$-\frac{59}{24}$	$\frac{37}{24}$	$-\frac{9}{24}$

The difficulty in obtaining the initial values for starting the integration can be overcome by computing the first n values of f by a single-step method, and then applying the multistep algorithm for the subsequent computations.

If the shorthand notation $f_k = f(kT, x_k, u_k)$ is introduced, the fourth-order algorithm reads

$$x_{k+1} = x_k + T \left[\frac{55}{24} f_k - \frac{59}{24} f_{k-1} + \frac{37}{24} f_{k-2} - \frac{9}{24} f_{k-3} \right] \quad (3.27)$$

This form looks very similar to the one found in equation 3.15 for computing the output of a general dynamic SISO system. The difference is that the new state vector is a linear combination not just of previous states, but of the nonlinear function of those previous states. A vector product notation similar to that of equation 3.15 can be as follows:

$$x_{k+1} = \beta^T f_{0n} + x_k \quad (3.28)$$

with the corresponding vectors

$$\beta = [\beta_0 \ ... \ \beta_n]^T, \qquad f_{0n} = [f_k \ ... \ f_{k-n}]^T \quad (3.29)$$

Thus the calculations to be performed for x_{k+1} are merely one evaluation of the vector function f for the purpose of obtaining the first element f_k of f_{0n} and the linear combination of the elements of f_{0n}.

3.6 Signal Processing Applications

Low-pass filters are used to avoid the effects of aliasing in discrete controllers when connected to a continuous plant. Normally, the filters are realized as analog systems and are regarded as a part of the plant. In some cases, mixed analog and digital techniques are applied where a part of the lowpass filter is produced as a discrete system. However, in most cases this approach offers no advantage or even yields worse performance than a purely analog design of the anti-aliasing filters (see [Brügge 89]). In particular, the oversampling method which is very popular in off-line signal processing (example: compact disc players) introduces substantially longer time delays than analog filters and thus has a negative influence on the stability of a control loop.

Anti-aliasing filters with linear phase response are desirable because they permit approximating the filter by a simple time delay. This eases significantly the task of designing the control loop.

It is often interesting or necessary to observe the spectrum of a signal, for example in order to know the noise level, to adapt parameters, or just for documentation purposes. An estimation method using the fast Fourier transform (FFT) is thus applied. The FFT and other methods for the spectrum estimation are therefore briefly described here. Many other signal transforms beyond the scope of this survey exist (see for example [Beauch 87], [Elliot 87]). They are used in special applications such as image processing or speech processing.

3.6.1 Digital Filtering

Discrete filters are produced by two basic approaches: either as IIR (Infinite Impulse Response) or as FIR (Finite Impulse Response) filters. The differences between these two approaches are outlined below.

Finite Impulse Response (FIR) Filters

FIR filters are characterized by the fact that their impulse response is limited in time. Therefore, the output u is expressed in terms of the input e only and reads

$$u_k = \sum_{i=0}^{q} b_{q-i} \, e_{k-i} \tag{3.30}$$

Reformulated as a vector product, this can be written as

$$u_k = b^T e_{0q} \tag{3.31}$$

with the corresponding vectors $b = [b_q \ ... \ b_0]^T$ and $e_{0q} = [e_k \ ... \ e_{k-q}]$ (cf. Section 3.3.1). This corresponds to a transfer function of

$$G(z) = \frac{U(z)}{E(z)} = b_q z^q + b_{q-1} z^{q-1} + ... + b_1 z + b_0 \tag{3.32}$$

Such a filter requires a large number of stages or taps to give a smooth amplitude response, typically from a few dozen to several hundred, which causes unwanted time delays in control systems. Apart from their guaranteed stability, the main advantage of this type of filter lies in the exact linear phase response. This allows the use of a simple delay model.

Many methods have been proposed for the design of FIR filters which aim at a smooth amplitude response. One well-known algorithm, the Remez exchange method, is described in [MCPaRa 73] and other solutions are presented in [RabGol 75], [CroRab 75], [Elliot 87], [KamWel 83], among others.

From a computational point of view an FIR filter is a simple system since it requires only the evaluation of the dot product 3.31 with a total of q multiplications and q-1 additions when computed in the time domain.

Infinite Impulse Response (IIR) Filters

IIR filters are the discrete equivalent to the classical continuous analog filter and incorporate a feedback of their past outputs. Therefore, in contrast to FIR filters, such filters can become unstable. With input e and output u, the difference equation is

$$u_k = \sum_{i=0}^{q} b_{q-i}\, e_{k-i} - \sum_{i=1}^{n} a_{n-i}\, u_{k-i} = b^T e_{0q} - a^T u_{1n} \qquad (3.33)$$

with

$$a = [a_{n-1} \ldots a_0]^T, \qquad b = [b_q \ldots b_0]^T$$

$$e_{0m} = [e_k \ldots e_{k-m}]^T, \qquad u_{1n} = [u_{k-1} \ldots u_{k-n}]^T$$

An FIR filter may be created by discretizing by an appropriate method a continuous filter designed with the desired characteristics. However, a trade-off is necessary between good amplitude response (small ripple, steep decrease for low-passes) and small and linear phase response. As pointed out in [Brügge 89], Bessel low-passes possess the most linear phase response. This is desirable for easy modelling of the filter by a time delay. The price for this is a relatively slow decrease in the amplitude response. Since an absolutely constant amplitude response in the pass-band is not necessary in control applications (unlike in pure filtering), it is more important to keep the phase delay as small as possible in order to gain stability in the loop.

Discretizing an analog design must be done carefully in order not to lose precious phase reserve or even the stability of the filter itself. It is recommended to use the bilinear transform (Tustin's approximation) with preceding prewarping (see [RabGol 75]). The preservation of stability is then guaranteed and a good amplitude response results.

Another approach is to design the filter in the z plane directly, either by pole placement or by approximating the frequency response at equidistant points (frequency sampling). This is an example of a method of design by optimization.

An abundant literature exists on the subject of designing IIR filters, from the classic tables by Zverev ([Zverev 67]), books by [RabGol 75] and [Elliot 87] to papers such as [LimLiu 88], [VlcUnb 89], and [SreAga 92].

IIR filters are easily evaluated in the time domain, since the computation of the next output requires the evaluation of the two dot products in 3.33 with a total of n+q multiplications and n+q-1 additions.

3.6.2 Signal Transforms

The Fourier transform is a classic. It was developed by Joseph Fourier at the beginning of the nineteenth century to compute the temperature distribution in an iron bar heated at one end. Many other transforms have been defined since, mostly for special purposes, e.g., for speech processing or image processing (see [Blahut 87], [Beauch 87], and [Elliot 87]). Recently, the focus has shifted to multi-dimensional signal processing.

A major problem with signal transforms has always been the computational effort required. In the case of the discrete Fourier transform, a great variety of transforms have been proposed, each differing in the number of computations. The twiddle-factor FFT approach has been selected here to show the kind of computations required for a discrete transform.

Discrete and Fast Fourier Transforms

The fast Fourier transform has become a widely used tool in engineering. Applications range from correlation analysis and filtering to spatial estimation in any area of engineering science (see [Brigha 88]). For image processing purposes, the two-dimensional Fourier transform is frequently used. Since its computational properties are equal to those of the one-dimensional Fourier transform, only the latter is discussed here.

The method of efficiently computing the discrete Fourier transform has become highly popular due to the paper by Cooley and Tukey [CooTuk 65], although Good had earlier proposed another algorithm for this task [Good 58]. In fact, Cooley and others later discovered that the first concepts of the FFT algorithm date back to C.F. Gauss in the seventeenth century [Cooley 92]. The computational effort for evaluating the complex N point Fourier series with the simplest radix-2 algorithm is basically $2 N \log N$ real multiplications and $3 N \log N$ real additions [Bergla 68], compared to $4 N^2$ and $2 N^2$, respectively, for the direct computation of the discrete Fourier transform. Unless noted differently, in this context "log" always denotes the logarithm to the base 2.

While the group of the Cooley-Tukey algorithms is very commonly used for its efficiency, another, less frequently employed but even more efficient method has been developed by Winograd [Winogr 78]. The Winograd group of the so-called prime factor algorithms uses the fact that the factors of N are mutually prime. Not only is the number of multiplications greatly reduced thereby but also the amount of additions needed decreases. This approach has been explored further by Temperton (cf. [Temper 83b], [Temper 85]) and Stasinski [Stasin 91]).

The more general case of N being composed of any factors has been investigated by Duhamel and Hollman ([DuhHol 84]) and by Temperton ([Temper 83a]). The latter regards the Fourier transform as the product of a vector containing the signal samples by the transform matrix. In his paper the various ways to factorize this matrix in order to isolate the single computational steps are summarized.

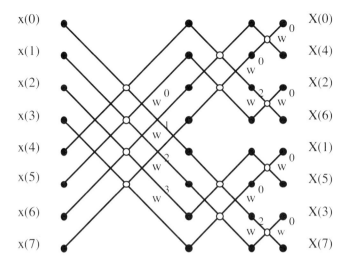

Figure 3.1 Eight-Point Radix-2 FFT Shown as Butterfly Diagram

Many variants of the FFT are subtle variations of the number theoretic mapping of the linear time space into the multidimensional index space of the transforms. For a recent review of the many approaches see [DuhVet 90].

The graphical description of the FFT by the "butterfly" diagrams is very popular, but not very useful for transforms of the lengths used in practice. However, they give a good insight into the structure of the algorithm. Figure 3.1 depicts an eight-point radix-2 fast Fourier transform.

Recursive Formulation of the One-dimensional Mixed-radix FFT

The ordinary discrete Fourier transform X(n) of a signal x(k) at times t = kT, where T stands for the sampling interval, is described for any number N of samples by the transform pair

$$X(n) = \sum_{k=0}^{N-1} x(k) \, e^{-j \frac{2\pi nk}{N}} \qquad (3.34)$$

and

$$x(k) = \frac{1}{N} \sum_{n=0}^{N-1} X(n) \, e^{j \frac{2\pi nk}{N}} \qquad (3.35)$$

For $N = r_1 r_2 .. r_m$, equation 3.34 can be transformed into a recursive formulation (see [Bergla 67]).

Using the abbreviation $w_N = e^{-\frac{2\pi j}{N}}$, the transform can conveniently be written as

$$X(n) = \sum_{k=0}^{N-1} x(k) \, w_N^{nk} \tag{3.36}$$

Since N is composed of m factors $r_1 r_2 .. r_m$, the variables n and k can be expressed as

$$n = n_{m-1}(r_1 r_2 .. r_{m-1}) + n_{m-2}(r_1 r_2 .. r_{m-2}) + .. + n_1 r_1 + n_0 \tag{3.37a}$$

and

$$k = k_{m-1}(r_2 r_3 .. r_m) + k_{m-2}(r_3 r_4 .. r_m) + .. + k_1 r_m + k_0 \; . \tag{3.37b}$$

The range of the respective coefficients n_{i-1} and k_i is

$$n_{i-1} = 0, 1, 2, .. \; r_i - 1 \qquad\qquad 1 \le i \le m \tag{3.38a}$$

$$k_i = 0, 1, 2, .. \; r_{m-i} - 1 \qquad\qquad 0 \le i \le m - 1 \tag{3.38b}$$

Thus a unique representation of n and k is possible for all values between 0 and N-1 with the sets $(n_{m-1}, n_{m-2}, .. , n_0)$ and $(k_{m-1}, k_{m-2}, .. , k_0)$, respectively. This allows 3.36 to be written as

$$X(n_{m-1}, n_{m-2}, .. , n_0) = \tag{3.39}$$

$$\sum_{k_0} \sum_{k_1} .. \sum_{k_{m-1}} x_0(k_{m-1}, k_{m-2}, .. , k_0) \, w_N^{nk}$$

with $x_0(k_{m-1}, k_{m-2}, .. , k_0) = x(k)$.

Rearranging the terms of the product nk yields

$$nk = n_0 k_{m-1}(r_2 r_3 .. r_m) + (n_1 r_1 + n_0) k_{m-2}(r_3 r_4 .. r_m) + .. + \tag{3.40}$$

$$+ (n_{m-1}(r_1 r_2 .. r_{m-1}) + .. + n_0) k_0$$

([Bergla 67], [Brigha 88]). Terms with $r_1 r_2 .. r_m = N$ can be omitted since $w_N^N = 1$.

The main advantage of this formulation is that each term of the sum contains only one factor k_i, such that the multiplications by w_N^{nk} can be distributed to the sums over the respective k_i. Therefore, intermediate variables can be defined, for example,

$$x_1(n_0, k_{m-2}, .., k_0) = \qquad (3.41)$$

$$\sum_{k_{m-1}} x_0(k_{m-1}, .., k_0) \, w_N^{n_0 k_{m-1}(r_2 .. r_m)}$$

and so forth, leading to the general formulation

$$x_i(n_0, n_1, .., n_{i-1}, k_{m-i-1}, .., k_0) =$$

$$\sum_{k_{m-i}} x_{i-1}(n_0, n_1, .., n_{i-2}, k_{m-i}, .., k_0) \cdot$$

$$w_N^{[n_{i-1}(r_1 r_2 .. r_{i-1}) + .. + n_0] k_{m-i}(r_{i+1} .. r_m)} \qquad (3.42)$$

$$i = 1, 2, .. m.$$

Equation 3.42 represents the recursive formulation of the Cooley-Tukey fast Fourier transform algorithm.

The final result is obtained by the relation

$$X(n_{m-1}, .., n_0) = x_m(n_0, .., n_{m-1}) \qquad (3.43)$$

The interchange of the order of the coefficients of n means an unscrambling of the results, the well-known bit reverse permutation for radix-2 FFT algorithms (where all r_is are 2, i.e., N is a power of two).

The product nk can also be factored as follows:

$$\begin{aligned} nk = \; & n_0 k_{m-1}(r_2 r_3 .. r_m) + n_0 k_{m-2}(r_3 r_4 .. r_m) + \qquad (3.44) \\ & + n_1 r_1 k_{m-2}(r_3 r_4 .. r_m) + \\ & + (n_1 r_1 + n_0) k_{m-3}(r_4 r_5 .. r_m) + .. + \\ & + n_{m-2}(r_1 r_2 .. r_{m-2}) k_1 r_m + \\ & + (n_{m-2}(r_1 r_2 .. r_{m-2}) + .. + n_0) k_0 + \\ & + n_{m-1}(r_1 r_2 .. r_{m-1}) k_0 + 0 \end{aligned}$$

By taking together two successive terms, the general expression for x_i becomes

$$x_i(n_0, n_1, ..., n_{i-1}, k_{m-i-1}, .., k_0) = \qquad (3.45)$$

$$\left[\sum_{k_{m-i}} x_{i-1}(n_0, n_1, .. , n_{i-2}, k_{m-i}, .. , k_0) \, w_N^{n_{i-1} k_{m-i} \frac{N}{r_i}} \right] \cdot$$

$$w_N^{[n_{i-1}(r_1 r_2 .. r_{i-1}) + .. + n_0] \, k_{m-i-1} (r_{i+2} .. r_m)}$$

$$i = 1, 2, .., m \text{ and } k_{-1} = n_{-1} = 0, r_{m+1} = r_{m+2} = 1.$$

Again, equation 3.43 applies to the final result.

In the intermediate stages, the values of x_i in equations 3.45 and 3.42 differ from each other, while the final result $X(n)$ must obviously be the same.

The last step 3.43 is an unscrambling of the results which is computationally expensive. Burrus has assembled ways ([Burrus 88]) to modify radix-2^m and mixed-radix FFT algorithms so that the results arrive in the same order as from radix-2 or radix-4 transforms. Simple unscramblers exist for these cases (see [BurPar 85]).

Each sum in brackets in 3.45 represents a r_i-point DFT, while the following term is called a "twiddle factor". It twiddles the phase of the whole partial DFT by multiplying the result of the transform by a complex factor with absolute value one. The advantage of this twiddle factor algorithm over the formulation in 3.42 lies in the fact that the r_i-point transform can be evaluated with a minimal number of multiplications. Regardless of the value of N, the factors in the sum are ± 1 for $r = 2$, ± 1 and $\pm i$ for $r = 4$, and $\pm 1, \pm i, \pm e^{\pm \frac{i\pi}{4}}$ for $r = 8$. For the single stages of the FFT, standard radix-r algorithms can thus be used and the intermediate results need to be modified only by the appropriate twiddle factor.

Radix-4 and radix-8 algorithms are the most efficient since the factors in the sum take on simple values. They require substantially fewer multiplications and additions than radix-2 algorithms (see [Bergla 68]).

The factorized formulation also very clearly exhibits the basic concept of the fast Fourier transform, namely the decomposition of a linear sequence of N points into m dimensions of length r_i.

As an illustration, the formulas for a 512-point transform in the radix-8 formulation will be given. As $512 = 8^3 = N = r_1 r_2 r_3$, the expressions to be evaluated are

$$x_1(n_0, k_1, k_0) = \left[\sum_{k_2 = 0}^{7} x_0(k_2, k_1, k_0) w_{512}^{n_0 k_2 \, 64} \right] w_{512}^{n_0 k_1 \, 8} \qquad (3.46)$$

$$x_2(n_0, n_1, k_0) = \left[\sum_{k_1 = 0}^{7} x_1(n_0, k_1, k_0) w_{512}^{n_1 k_1 \, 64} \right] w_{512}^{(8n_1 + n_0) k_0}$$

$$x_3(n_0, n_1, n_2) = \left[\sum_{k_0 = 0}^{7} x_2(n_0, n_1, k_0) w_{512}^{n_2 k_0 \, 64} \right] w_{512}^{0}$$

$$X(n_2, n_1, n_0) = x_3(n_0, n_1, n_2)$$

with $w_{512} = e^{-\frac{2\pi j}{512}}$.

Matrix Formulation of the FFT

As indicated earlier, the discrete Fourier transform can also be described in terms of a matrix-vector product. Thus, equation 3.34 can be rewritten as

$$X = T_0 \, x \tag{3.47}$$

$$X = [X(0), X(1), .. X(N)]^T, \quad x = [x(0), x(1), .. x(N)]^T$$

where X and x are the vectors of the N frequency and time samples, respectively. T_0 denotes the N × N transform matrix.

It is inefficient to evaluate the product in this form, since it requires N^2 complex multiplications. But the transform matrix can be factorized in a product of sparse matrices, which allows computation of the transform with much less expense than previously. The approach in this notation is due to Pease [Pease 68]. However, his formulation of the FFT will not be repeated here. Instead, the factored matrix version of the twiddle-factor Cooley-Tukey algorithm will be given.

In equation 3.39, the splitting of the sum into m smaller sums has been introduced for N composed of m arbitrary factors, i.e., $N = r_1 r_2 .. r_m$. The introduction of the intermediate vectors x_i for the computation stage i in equation 3.45 suggests a factorization of T_0 in the form

$$x_1 = T_1 \, x_0, \quad x_2 = T_2 \, x_1, \quad .. \, , \quad x_m = T_m \, x_{m-1} \tag{3.48}$$

giving for x_m

$$x_m = \prod_{i=1}^{m} T_i \, x_{i-1} \tag{3.49}$$

X is obtained by unscrambling x_m by a permutation matrix which will be called P_{m+1}. This leads to

$$X = P_{m+1} \, x_m \tag{3.50}$$

According to equation 3.45, for each x_i the summation is over k_{m-i}, which represents a scrambled order of x_i as it is summed. After the summation, the result is multiplied by the twiddle factor which does not depend on the summation index.

This leads to the idea of partitioning the original transformation matrix for stage i of the computation as

$$T_i = P_i^T E_i C_i P_i \qquad (3.51)$$

where
- P_i performs the permutation to scrambled order
- C_i performs the basic r_i point transform at stage i
- E_i performs the multiplication by the twiddle factors
- P_i^T performs the inverse permutation to natural order

The permutation matrix P_i is filled mainly with zeros and with N ones. C_i is block-diagonal in structure, and E_i is strictly diagonal.

The main idea of the decomposition is twofold. On the one hand, the large blocks of similar computations on the data in a specific order (the r_i point transforms) are isolated by partitioning the whole fast Fourier transform into m stages. Each stage corresponds to one factor of N. On the other hand, each stage divides in single steps: the permutation of data, the N/r_i parallel r_i point transforms, the multiplication of the data vector by the twiddle factors, and unscrambling the data. The regular structure of the algorithm appears very clearly. One other basic property of the fast Fourier transform is that it is not data dependent.

For practical purposes, the factorization of T_i in 3.51 can be modified slightly by combining the inverse permutation at stage i-1 with the (different) permutation at stage i, yielding

$$T_i = E_i C_i P_{i,i-1} \qquad (3.52)$$

with $P_{i,i-1} = P_i P_{i-1}^T$, $i = 1, 2, \ldots, m$, and $P_0 = I_N$

For the specification of the matrices P, C, and E, a notation similar to that of Pease ([Pease 68]) will be used. However, the Kronecker product is defined differently here, but in accordance with the definition used by Temperton [Temper 83a].

More recently, this kind of description of the FFT algorithms has been compiled under the name Tensor Product formulation ([GrCoTo 92]).

The Kronecker product of two matrices A and B is defined by

$$A \times B = (a_{ij} B)$$

C_i thus results in the following:

$$C_i = I_{r_i}^N \times W_{r_i} \tag{3.53}$$

where

I_s denotes the identity matrix of dimension s.
$W_{r_i}(i, j) = w_{r_i}{}^{ij}$ stands for the square r_i point transform matrix, and
w_{r_i} is as defined in equation 3.36.

Therefore, C_i has a blockdiagonal form with $\dfrac{N}{r_i}$ square r_i point transform matrices W_{r_i} on the diagonal.

For the diagonal twiddle-factor matrix E_i, the expression with exponents is more convenient than one written in full powers of w_{r_i}. Therefore, any figure in the matrix will denote the exponent only, whereas all nondiagonal entries are written as period, meaning the value zero rather than the exponent zero.

Let $R_{r_i} = \text{diag}(0, 1, 2, \ldots, r_i-1)$. Then, by inspection of the twiddle factor, the last term in 3.45, E_i results as

$$E_i = \left[(I_{\frac{N}{r_i}}^N \times R_{r_i}) \, r_0 r_1 r_2 .. r_{i-1} + \right. \tag{3.54}$$

$$\sum_{k=2}^{i} (I_{r_0 r_1 .. r_{i-k}} \times R_{r_{i-k+1}} \times I_{r_{i-k+2} r_{i-k+3} .. r_m}) \, r_0 r_1 .. r_{i-k} \Big] \cdot$$

$$\left[I_{r_0 r_1 .. r_{i-1}} \times R_{r_{i+1}} \times I_{r_i r_{i+2} r_{i+3} .. r_m} \right] r_{i+2} r_{i+3} .. r_m$$

$i = 1, 2, \ldots, m-1$, and $r_0 = 1$, $E_m = I_N$

When comparing the above formula with the original formulation in equation 3.45, the first expression corresponds to the factor n_{i-1} of the exponent of w_N, the sum corresponds to the factors $n_{i-2} .. n_0$, and the last term in brackets corresponds to k_{m-i-1}.

Describing P_i requires a construction rule as follows. Any $x_i(l)$ is numbered according to

$$l = l_{m-1}(r_2 r_3 .. r_m) + l_{m-2}(r_3 r_4 .. r_m) + .. + l_1 r_m + l_0 \tag{3.55}$$

(see equation 3.37) with the ordered set $L = \{l_{m-1}, l_{m-2}, \ldots, l_1, l_0\}$, where l_{m-t} stands for the n or k at position t in 3.45.

For P_i form

$$L' = \{l_{m-1}, l_{m-2}, .. \, l_{m-i+1}, l_{m-i-1} \ldots, l_1, l_0\} \tag{3.56}$$

by omitting l_{m-i}. For all values of the respective l_{m-t} evaluate 3.55 without the term for l_{m-i} and order the corresponding values in ascending order. This will result in $\frac{N}{r_i}$ values for l'. Then a new element k_{m-i} is added to L', yielding

$$L'' = \{l_{m-1}, l_{m-2}, .. l_{m-i+1}, k_{m-i}, l_{m-i-1} .. , l_1, l_0\} \qquad (3.57)$$

where $k_{m-i} = 0, 1, .. , r_i -1$. Evaluated in the same order as L', L" denotes the order of $x_{i-1}(l'')$.

Therefore, for P_i, the columns are numbered from left to right according to l, i.e., in normal order, while the rows are numbered from top to bottom according to l", i.e., in scrambled order. The position where l and l" have the same value is marked by a 1, and all the other elements of P_i are zero.

When constructing $P_{i,i-1} = P_i P_{i-1}^T$, the columns are numbered according to l_{i-1}'' and the rows according to l_i''. Written in compact notation, this reads

$$P_{i,i-1} = (\delta_i'' l_{i-1}'') \qquad (3.58)$$

In sum, the fast Fourier transform is an algorithm which requires large computational effort, but due to its highly regular structure, the transform matrix can be divided into successive stages of permutations, small transforms, and the weighting of each intermediate value.

Convolution, Correlation, and Spectral Estimation

Convolution is applied for the computation of the output of a linear system, for example, given the input signal and the system's impulse response. The discrete convolution differs from its continuous equivalent in that it is periodic. Therefore it is also called circular convolution. If the sequences to be convolved have length N and are continued periodically then the resulting convolution is of length 2N and periodic.

Given a system impulse response h(k) and an input sequence x(k), both of length N and continued periodically, the system output y(k) is computed as follows:

$$y(k) = \sum_{l=0}^{N-1} x(l) h(k-l), \qquad 0 \le k \le N-1 \qquad (3.59)$$

The same operation can be performed in the frequency domain with the discrete Fourier transforms of the signals, reading

$$Y(n) = X(n) H(n), \qquad 0 \le n \le N-1 \qquad (3.60)$$

In the frequency domain, only the point-wise multiplication of the transformed signals is required, rather than the scalar product of x and h shifted by k points. Therefore, it is often attractive to perform the operation in the frequency domain, especially if the result has to be transformed to this domain in any case. An efficient algorithm for the discrete Fourier transform must be available in order to do this quickly.

The cross-correlation and the power-density spectrum of signals are useful tools for gaining information in radar, adaptive filtering, seismic data processing, speech processing, underwater acoustics, and many other applications. For practical purposes it is generally assumed that the signals are stationary, because they have to be observed over a period of time before they can be processed.

The discrete correlation of two signals x and y of length N is defined by

$$r_{xy}(k) = \sum_{l=0}^{N-1} x(l)\, y(k+l), \qquad 0 \leq k \leq N-1 \qquad (3.61)$$

When transforming both signals to X(n) and Y(n), the transformed cross-correlation function becomes

$$R_{xy}(n) = Y(n)\, X^*(n) \qquad (3.62)$$

where * denotes complex conjugate. This means that in the frequency domain the correlation changes into a point-by-point multiplication of the transformed signals.

From continuous signal processing, it is known that the cross-power density function $S_{xy}(k)$ and the crosscorrelation $r_{xy}(k)$ are a Fourier pair. However, in the discrete case all spectral functions are inaccurate to a certain degree, since the observations of the signals cover only a limited period of time.

In order to reduce the error, one method is to average a number of estimates $\hat{S}_{xy}(k)$. This was first described by Blackman and Tukey [BlaTuk 59] and has been further refined by Welch [Welch 67].

The basic idea is to partition the original sequences x(k) and y(k) of length N into k segments of length L, possibly overlapping. These segments are weighted with a windowing function (cf. [Harris 78], [Adams 91]) and subsequently transformed. The transformed segments are then summed and averaged.

From a computational point of view the estimation of spectra is an application of the fast Fourier transform, with additional averaging at the end, consisting of additions and divisions on an array of data of length L.

3.7 Matrix Computations

The concept of matrix notation is very attractive for the formulation of mathematical problems. On the one hand, matrices and vectors allow a very compact and concise notation for many applications, and on the other, vector computers offer efficient ways to solve systems of linear equation in a short time. This has made it attractive to reformulate differential equations, for example for hydrodynamic problems, in large linear equation systems, which in turn are solved efficiently on vector computers.

With the introduction of parallel computers, increased emphasis has been put on exploiting aspects other than vectorization of linear equation systems. However, most of the commercial systems still consist of a very limited number of processors, which in turn are heavily pipelined vector machines. Nevertheless, the aspect of parallelizing computations will gain importance. Many approaches have already been made, as described in [Ortega 88] or [Modi 88].

From the discussion of the algorithms for control in Sections 3.3 and 3.4, and from the algorithms outlined in Section 3.6 for signal processing, the significance of the matrix and vector operations has become evident. Most of the algorithms treated can be formulated in vector form.

After outlining the computational effort for the basic operations with vectors and matrices, including block matrix techniques, further algorithms are covered here which may be needed in more advanced applications but which will be rarely used in real-time environments. The solution of linear systems and matrix inversion will be briefly outlined. The eigenvalue/eigenvector decomposition and computation of the singular value decomposition of matrices, however, will not be treated, because they require fairly subtle numerics and are generally too expensive to be computed in real-time.

Also, the computation of the matrix exponential is not considered here. While it is an important step in the design of controllers, it should be avoided for run-time computations due to its high demands on computational effort and numerical accuracy (cf. [MolVanL 78]). Many implementation aspects of matrix algorithms are outlined in [VanLoa 90].

Generally, all matrices and vectors are assumed to have real-valued elements. If complex numbers appear, as with the Fourier transform, all algorithms apply in the same way, except that the ordinary transposition has to be replaced by the conjugate transposition, where the transposed elements are replaced by their complex conjugates.

The notation will vary with requirements. The computations are expressed either in terms of whole vectors (e.g., the addition of such n-dimensional vectors as $z = x + y$), or as single components of the vectors or matrices, such as $z_i = x_i + y_i$, $1 \leq i \leq n$. Vectors will be regarded as columns, otherwise they are affixed with the transpose sign "T". Matrix names consist of capital letters, while single elements are referred to by the same letter in lower case, with additional indices indicating the position. Capital letters with indices are used for submatrices. Many of the algorithms and much of this notation are referenced in [GolVanL 89].

3.7.1 Basic Vector and Matrix Computations

The basic operations with vectors and matrices comprise addition and multiplication of operands of corresponding dimensions. This forms the main application of matrices and vectors in real-time applications for signal processing and digital control. In the scientific computing community, a larger collection of Fortran subroutines for this purpose has become known as BLAS (Basic Linear Algebra Subroutines). These subroutines serve for broader computations than are needed in real-time digital control.

An older subroutine collection is LINPACK, which incorporates most of the basic algorithms.

Historically, three levels of the BLAS code have emerged. Starting with level one, only operations on vectors were covered ([LaHaKi 79]). Then, matrix-vector operations were soon included, typically matrix times vector multiplications ([DoDCHa 88]). These subroutines have been implemented on many vector computers and are the basis of the highly efficient software packages for solving linear equation systems. Level three of BLAS ([DoDCHa 87]) incorporates operations on the matrix-matrix level and will gain even more importance in the future.

Scalar-vector Computations

There is only one operation which incorporates a scalar α and a vector x of dimension n. This is the scalar-vector multiplication $z = \alpha x$ ($z_i = \alpha x_i$) requiring n multiplications.

Vector-vector Computations

The most frequent operations among vectors in control and signal processing are the inner or dot product, sometimes referred to as scalar product, and the addition of two vectors. The inner product appears whenever difference equations have to be evaluated, for example with the general SISO controller 3.15, or with FIR and IIR filters, as described in equations 3.30 and 3.33, and in context with correlation (equation. 3.59). Expressed by the vectors x and y of dimension n, it reads

$$c = x^T y = \sum_{i=1}^{n} x_i y_i \tag{3.63}$$

The computational expense is n multiplications and n-1 additions. The outer product of two vectors x and y of dimensions m and n, respectively, yielding a m × n matrix Z is defined by

$$Z = x y^T, \qquad z_{ij} = x_i y_j, \qquad 1 \le i \le m, \ 1 \le j \le n \tag{3.64}$$

However, this type of vector product is rarely encountered in signal processing. The number of multiplications to be performed is nm.

On vector computers, the term vector multiplication of two vectors of the same dimension means the following, requiring n multiplications:

$$z = x \times y, \qquad z_i = x_i y_i \qquad 1 \le i \le n \tag{3.65}$$

The addition of two vectors x and y yielding the vector z is described by

$$z = x + y, \qquad z_i = x_i + y_i \tag{3.66}$$

This addition requires n additions.

A well-known composition of scalar-vector multiplication and vector addition, though less frequently used in the applications discussed here, is the SAXPY operation (scalar alpha (times) x plus y, while in the original BLAS definition S represents single precision). It is denoted by

$$z = \alpha x + y, \qquad z_i = \alpha x_i + y_i \qquad (3.67)$$

Because it contains a scalar-vector multiplication and the addition of two vectors, n multiplications and additions have to be computed.

Matrix-vector Computations

Matrix-vector computations, specifically the multiplication of a matrix by a vector, are among the central operations in control and signal processing when systems are expressed in state space. They appear with linear systems (equations 3.3 and 3.4), state feedback (equation. 3.16), and the fast Fourier transform 3.49, to name a few.

The so-called row version of the matrix-vector multiplication is to see the result vector z composed of single dot products. If matrix A has m rows and n columns, the following algorithm results:

$$z = A x, \qquad z_i = \sum_{j=1}^{n} a_{ij} x_j, \qquad 1 \le i \le m, \ 1 \le j \le n \qquad (3.68)$$

However, it is possible to express z as the sum of the columns of A multiplied by the corresponding element of x, giving the column version of the matrix-vector multiplication. Introducing the n intermediate vectors $_jt$, the result z can be expressed as

$$_jt_i = x_j a_{ij} \qquad 1 \le j \le n$$

$$z_i = \sum_{j=1}^{n} {_jt_i} \qquad 1 \le i \le m \qquad (3.69)$$

Both variants require mn multiplications, and m(n-1) additions.

The GAXPY operation appears occasionally. It is defined by the equation

$$z = y + Ax \qquad (3.70)$$

with the vectors y and z of dimension m, the vector x of dimension n and the m × n matrix A. The number of additions and multiplications is mn.

Scalar-matrix Computations

As in the scalar-vector case, the only useful operation involving a scalar α and the $m \times n$ matrices A and B is the product defined by

$$B = \alpha A, \qquad b_{ij} = \alpha\, a_{ij}, \qquad 1 \leq i \leq m,\ 1 \leq j \leq n \qquad (3.71)$$

which can be computed element by element.

Matrix-matrix Computations

The addition (or subtraction) of two matrices A and B of the same dimensions $m \times n$ yields the result

$$C = A + B, \qquad c_{ij} = a_{ij} + b_{ij}, \qquad 1 \leq i \leq m,\ 1 \leq j \leq n \qquad (3.72)$$

Given an $m \times n$ matrix A and an $n \times m$ matrix B, the resulting square matrix C of dimension $m \times m$ is given by

$$C = A\,B, \qquad c_{ij} = \sum_{k=1}^{n} a_{ik}\, b_{kj}, \qquad 1 \leq i \leq m,\ 1 \leq j \leq m \qquad (3.73)$$

This formulation relies on the computation of the elements of C by inner products as defined in 3.63. It is also referenced as the ijk form of matrix multiplication, named after the order of the indices. However, it is possible to compute the SAXPY operation in the loop in two variants as given in 3.68 and 3.69. If this innermost loop runs over either i or j instead of k, another three possibilities for the computations arise. This gives a total of six possible ways to compute the product of two matrices. The different variants and their impact on the performance of the implementation are discussed in further detail in [DoGuKa 84].

Table 3.4 shows six possibilities for the arrangement of the loops.

Table 3.4 *Loop Ordering for Matrix Multiplication*

Index Order	Inner Loop	Middle Loop
ijk	Inner Product	Vector × Matrix
jik	Inner Product	Matrix × Vector
ikj	SAXPY	GAXPY by Rows
jki	SAXPY	GAXPY by Columns
kij	SAXPY	Outer Prod. (rows)
kji	SAXPY	Outer Prod. (columns)

The efficiency of a specific implementation of the matrix multiplication depends very much on the data access scheme provided by the computer and by the programming language.

The basic computational effort is nm^2 multiplications and $(n-1)m^2$ additions. This high complexity explains why it is worth exploiting every known structural property of the matrices, for example diagonal sparseness.

Block Matrix Computations

With the advent of computers incorporating multiple processors it has become attractive to distribute the computations over the processing elements. For this purpose, the block matrix notation is very appropriate, since it allows the expression of algorithms in terms of matrix elements which are matrices themselves. The computational rules, however, remain the same as for the calculation on the level of matrix elements. It must be kept in mind that matrix products normally do not commute, in contrast to scalar multiplications.

For example, a vector x and a matrix A could be partitioned as follows:

$$x = \begin{bmatrix} x_1 \\ x_2 \\ \ldots \\ x_p \end{bmatrix}, \qquad A = \begin{bmatrix} A_{11} & A_{12} & \ldots & A_{1p} \\ A_{21} & A_{22} & \ldots & A_{2p} \\ \ldots & \ldots & \ldots & \ldots \\ A_{q1} & A_{q2} & \ldots & A_{qp} \end{bmatrix} \qquad (3.74)$$

The subvectors x_i and the submatrices A_{ij} can have any matching dimensions. The vector $y = Ax$ can then be expressed as

$$y = \begin{bmatrix} y_1 \\ y_2 \\ \ldots \\ y_p \end{bmatrix}, \qquad y_i = \sum_{j=1}^{q} A_{ij} x_j \qquad 1 \leq j \leq q \qquad (3.75)$$

Obviously, it is possible to formulate all the computations described in previous sections in the same manner, for example the solution of systems of linear equations.

For multiprocessor systems the introduction of block matrix algebra is a promising approach, since the costs of transferring submatrices and the results are compensated by the gain in parallelism. For this type of arithmetic a collection of standard subroutines is under development ([DeDoDC 87]) and is called LAPACK (Linear Algebra PACKage).

3.7.2 Solution of Systems of Linear Equations

Given a system of linear equations

$$Ax = b \qquad \text{with } A \in R^{m \times n}, \ x \in R^{1 \times n}, \ b \in R^{1 \times m} \qquad (3.76)$$

the solution x for a given matrix A and vector b is sought.

If only the case of square matrices A is considered (i.e., m = n), then for a nonsingular A a solution can be found. One favourite method for determining x is the Gaussian elimination using LU decomposition. There the matrix A is decomposed so that with the intermediate vector w the following equations result:

$$A = L R, \qquad Lw = b, \qquad Ux = w \qquad (3.77)$$

where L is a lower triangular matrix and U an upper triangular matrix. The vector w is determined by forward substitution, and the final solution x results from backward substitution.

This decomposition requires approximately $\frac{n^3}{3}$ additions and multiplications. If A is Hermitian, i.e., if the equations are positive definite, the Cholesky decomposition $A = R^T R$ can be applied instead of the LU decomposition, thus halving the computational effort.

There exists an algorithm by Strassen ([Strass 69]) where the effort is proportional to less than the third power of n, namely only to $n^{\log_2 7}$ which is approximately $n^{2.8}$.

The crucial point to obtain numerically exact results is a good choice of pivots. Apart from scaling the equations to get coefficients of similar order of magnitude, it is essential to choose pivots that are as large as possible to reduce roundoff errors. Therefore, in a routine to solve general problems a flexible pivoting strategy has to be implemented.

In on-line applications for control and digital signal processing the solution of linear systems is rarely computed, except when filters or compensators have to be tuned in real-time due to varying plant or excitation parameters. One case is the Kalman filter (equations 3.19-3.23) where the gain matrix L_k changes with each step.

3.7.3 Matrix Inversion

If the inverse A^{-1} of a square matrix A ($A^{-1}A=I$) is not explicitly needed its computation is avoided if possible. In an application of the Kalman filter, however, computation is inevitable.

The most straightforward way to do this is to use the exchange algorithm for the matrix to be inverted. The result, however, is n^3 additions and multiplications. It is more appropriate to compute the LR decomposition of A once and then to solve the n linear equation systems

$$A x_i = e_i \qquad e_{ik} = \partial(k) \text{ (Kronecker function)}, 1 \le i \le n \qquad (3.78)$$

Each solution x_i ($1 \le i \le n$) is the corresponding column of $A^{-1} = [x_1, x_2, ..., x_n]$ because it represents the vector orthogonal to the i-th row of A.

The decomposition needs $\frac{2n^3}{3}$ operations and is done only once. The subsequent solution of the n equations by forward and backward substitution requires an effort only proportional to n^2.

3.8 References

[Adams 91] J.W. Adams, "A New Optimal Window," *IEEE Trans. ASSP*, vol. 39, no. 8, pp. 1753-1769, 1991.

[Allen 85] J. Allen, "Computing Architecture for Digital Signal Processing," *Proceedings of the IEEE*, vol. 73, no. 5, pp. 852-873, 1985.

[ÅstWit 90] K. Åström and B. Wittenmark, *Computer-Controlled Systems: Theory and Design*. (2nd edition), London: Prentice-Hall International, Inc., 1990.

[Beauch 87] K. Beauchamp, *Transforms for Engineers: A Guide to Signal Processing*. Oxford: Clarendon Press, 1987.

[Bergla 67] G.D. Bergland, "The Fast Fourier Transform Recursive Equations for Arbitrary Length Records," *Mathematics of Computation*, vol. 21, pp. 236-238, 1967.

[Bergla 68] G.D. Bergland, "A Fast Fourier Transform Algorithm Using Base 8 Iterations," *Mathematics of Computation*, vol. 22, pp. 275-279, 1968.

[BlaTuk 59] R.B. Blackman and J.W. Tukey, *The Measurement of Power Spectra From the Point of View of Communications Engineering*. New York: Dover Publications, Inc., 1959. Republished from the January and March issues of *The Bell System Technical Journal*, vol. 37, 1958.

[Blahut 87] R.E. Blahut, *Fast Algorithms for Digital Signal Processing*. Reading, MA, a.o.: Addison-Wesley Publishing Co., 1987.

[Brigha 88] E.O. Brigham, *The Fast Fourier Transform and its Applications*. London: Prentice-Hall International Inc., 1988.

[Brügge 89] T. Brüggemann, "Anti-Aliasing-Filterung in regeltechnischen Anwendungen," Report No. 23, Swiss Federal Institute of Technology, Zurich (ETH Zurich), Measurement and Control Laboratory, Sept. 1989.

[ByrHin 75] G.D. Byrne and A.C. Hindmarsh, "A Polyalgorithm for the Numerical Solution of Ordinary Differential Equations," *ACM Trans. on Mathem. Software*, vol. 1, no. 1, pp. 71-96, 1975.

[Cooley 92] J.W. Cooley, "How the FFT Gained Acceptance", *IEEE Signal Processing Magazine*, vol. 9, no. 1, pp. 10-13, 1992. Reprinted from: *Proceedings of the ACM Conference on the History of Scientific and Numeric Computation*, May 1987, and from *A History of Scientific Computing*, Boston, MA: ACM Press, 1990.

[CooTuk 65] J.W. Cooley and J.W. Tukey, "An Algorithm for the Machine Calculation of Complex Fourier Series," *Mathematics of Computation*, vol. 19, pp. 297-301, 1965.

[CroRab 75] R. Crochiere and L.R. Rabiner, "Optimum FIR Digital Filter Implementations for Decimation, Interpolation, and Narrow-Band Filtering," *IEEE Trans. ASSP*, vol. 23, no. 5, pp. 444-456, 1983.

[DeDoDC 87] J. Demmel, J. Dongarra, J. Du Croz, *et al.*, "Prospectus for the Development of a Linear Algebra Library for High-Performance Computers," Report ANL/MCS-TM-97, Argonne National Laboratory, Argonne, IL, September 1987.

[DoDCHa 87] J.J. Dongarra, J. Du Croz, S. Hammarling, *et al.*, "A Proposal for a Set of Level 3 BLAS Basic Linear Algebra Subprograms," Report ANL/MCS-TM-88, Argonne National Laboratory, Argonne, IL, April 1987.

[DoDCHa 88] J.J. Dongarra, J. Du Croz, S. Hammarling, *et al.*, "An Extended Set of Fortran Basic Linear Algebra Subprograms," *ACM Trans. on Mathem. Software*, vol. 14, no. 1, pp. 1-17, 1988.

[DoGuKa 84] J.J. Dongarra, F.G. Gustavson and A. Karp, "Implementing Linear Algebra Algorithms for Dense Matrices on a Vector Pipeline Machine," *SIAM Review*, vol. 26, no. 1, pp. 91-112, 1984.

[DuhHol 84] P. Duhamel and H. Hollmann, "'Split Radix' FFT Algorithm," *Electronics Letters*, vol. 20, no. 1, pp. 14-16, 1984.

[DuhVet 90] P. Duhamel and M. Vetterli, "Fast Fourier Transforms: A Tutorial Review and a State of the Art," *Signal Processing*, vol. 19, no. 4, pp. 259-299, 1990.

[Elliot 87] D.F. Elliott (ed.), *Handbook of Digital Signal Processing: Engineering Applications*. San Diego, New York, a.o.: Academic Press, Inc., 1987.

[FraPow 90] G. Franklin and J. Powell, *Digital Control of Dynamic Systems*, (2nd edition). Reading, MA, a.o.: Addison-Wesley Publishing Co., 1990.

[Gear 71] C.W. Gear, "Algorithm 407 – DIFSUB for Solution of Ordinary Differential Equations," *Communications of the ACM*, vol. 14, no. 3, pp. 185-190, 1971.

[Geerin 86] H.P. Geering, "Entwurf robuster Regler mit Hilfe von Singulärwerten; Anwendung auf Automobilmotoren," GMA-Bericht 11, pp. 125-145, 1986.

[Geerin 87] H.P. Geering, "Neuere Methoden für den Entwurf robuster Regler," *Bulletin SEV*, vol. 78, no. 7, pp. 346-349, 1987.

[Goldbe 91] D. Goldberg, "What Every Compouter Scientist Should Know About Floating-Point Arithmetic," *ACM Computing Surveys*, vol. 23, no. 1, pp. 5-48, 1991.

[GolVanL 89] G. Golub and C. Van Loan, *Matrix Computations*, (2nd edition). Baltimore, London: The Johns Hopkins University Press, 1989.

[Good 58] I.J. Good, "The Interaction Algorithm and Practical Fourier Analysis," *J. Royal Statist. Soc., Ser. B*, vol. 20, pp. 361-375, 1958. Addendum in vol. 22, pp. 372-375, 1960.

[GuSaTi 85] G.K. Gupta, R. Sacks-Davis and P.E. Tischer, "A Review of Recent Developments in Solving ODEs," *ACM Computing Surveys*, vol. 17, no. 1, pp. 5-47, 1985.

[GrCoTo 92] J. Granata, M. Conner and R. Tolimieri, "The Tensor Product: A Mathematical Programming Language for FFTs and other Fast DSP Operations," *IEEE Signal Processing Magazine*, vol. 9, no. 1, pp. 40-48, 1992.

[HaNøWa 87] E. Hairer, S. Nørsett and G. Wanner, *Solving Ordinary Differential Equations I: Nonstiff Problems*. Berlin, Heidelberg, New York: Springer-Verlag, 1987.

[HaiWan 91] E. Hairer and G. Wanner, *Solving Ordinary Differential Equations II: Stiff and Differential-Algebraic Problems*, Berlin, Heidelberg, New York: Springer-Verlag, 1991.

[Hansel 87] H. Hanselmann, "Implementation of Digital Controllers – A Survey," *Automatica*, vol. 23, no. 1, pp. 7-32, 1987.

[Harris 78] F.J. Harris, "On the Use of Windows for Harmonic Analysis with the Discrete Fourier Transform," *Proceedings of the IEEE*, vol. 66, no. 1, pp. 51-83, 1978.

[Hindma 74] A.C. Hindmarsh, "GEAR: Ordinary Differential Equation System Solver," Rep. UICD-30001, Revision 3, Lawrence Livermore National Laboratory, Livermore, CA, 1974.

[Hindma 80] A.C. Hindmarsh, "LSODE and LSODI, Two New Initial Value Ordinary Differential Equation Solvers," *ACM SIGNUM Newsletters*, vol. 15, pp. 10-11, 1980.

[HuEnJa 76] T.E. Hull, W.H. Enright and K.R. Jackson, "A User's Guide for DVERK – A Subroutine for Solving Non-Stiff ODEs," Tech. Rep. 100, Dept. of Computer Science, University of Toronto, Ont., 1976.

[Kalman 60] R.E. Kalman, "A New Approach to Linear Filtering and Prediction Problems," *Trans. ASME, Series D, J. Basic Eng.*, vol. 82, pp. 35-45, 1960.

[KamWel 83] Y. Kamp and C. Wellekens, "Optimal Design of Minimum Phase FIR Filters," *IEEE Trans. ASSP*, vol. 31, no. 4, pp. 922-926, 1983.

[Krogh 69]   F.T. Krogh, "A Variable Step, Variable Order Multistep Method for the Numerical Solution of ODEs," in *Information Processing 68*, A.J.H. Morrel (ed.). Amsterdam: North-Holland Publishing Co., 1969, pp. 194-199.

[KwaSiv 72]   H. Kwakernaak and R. Sivan, *Linear Optimal Control Systems*. New York, London, Sydney: Wiley-Interscience, 1972.

[LaHaKi 79]   C.L. Lawson, R.J. Hanson, D.R. Kincaid, *et al.*, "Basic Linear Algebra Subprograms for Fortran Usage," *ACM Trans. on Mathem. Software*, vol. 5, no. 3, pp. 308-323, 1979.

[LimLiu 88]   Y.C. Lim and B. Liu, "Design of Cascade Form FIR Filters with Discrete Valued Coefficients," *IEEE Trans. ASSP*, vol. 36, no. 11, pp. 1735-1739, 1988.

[MCPaRa 73]   J. H. McClellan, T.W. Parks and L.R. Rabiner, "A Computer Program for Designing Optimum FIR Linear Phase Digital Filters," *IEEE Trans. AU*, vol. 21, no. 6, pp. 506-526, 1973.

[Mendel 71]   J. M. Mendel, "Computational Requirements for a Discrete Kalman Filter," *IEEE Trans. Automatic Control*, vol. 16, no. 6, pp. 718-758, 1971.

[Modi 88]   J.J. Modi, *Parallel Algorithms and Matrix Computation*. Oxford Applied Mathematics and Computing Series, Oxford: Clarendon Press, 1988.

[MolVanL 78]   C. Moler and C. Van Loan, "Nineteen Dubious Ways to Compute the Exponential of a Matrix," *SIAM Review*, vol. 20, no. 4, pp. 801-836, 1978.

[Morone 83]   P. Moroney, *Issues in the Implementation of Digital Feedback Compensators*, MIT Press Series in Signal Processing, Optimization and Control, Alan S. Willsky (ed.). Cambridge, MA; London: MIT Press, 1983.

[Ortega 88]   J.M. Ortega, *Introduction to Parallel and Vector Solution of Linear Systems*, Frontiers in Computer Science Series, Arnold L. Rosenberg (ed.). New York, London: Plenum Press, 1988.

[Pease 68]   M.C. Pease, "An Adaptation of the Fast Fourier Transform for Parallel Processing," *Journal of the ACM*, vol. 15, no. 2, pp. 252-264, 1968.

[RabGol 75]   L.R. Rabiner and B. Gold, *Theory and Application of Digital Signal Processing*. Englewood Cliffs, NJ: Prentice-Hall, Inc., 1975.

[Rao 85]   S. Rao, *Regular Iterative Algorithms and their Implementation on Processor Arrays*. Ph.D. Dissertation, Stanford University, 1985. Diss. Abstracts No. DA8608214.

[RaoKai 87]   S. Rao and T. Kailath, "Architecture Design for Regular Iterative Algorithms," in *Systolic Signal Processing*, Earl E. Swartzlander (ed.). New York, Basel: Marcel Dekker, Inc., 1987, pp. 209-297.

[Schwar 86]   H. Schwarz, *Numerische Mathematik*. Stuttgart: B. G. Teubner, 1986.

[Sedgwi 73]   A.E. Sedgwick, "An Effective Variable Order Variable Stepsize Adams Method," Technical Report 53, Dept. of Computer Science, University of Toronto, Ont., 1973.

[Shampi 83]   L.F. Shampine, "Measuring Stiffness," Technical Report SAND83-1119, Sandia Laboratories, Albuquerque, NM, June 1983.

[ShaGea 79]   L.F. Shampine and C.W. Gear, "A User's View of Stiff Ordinary Differential Equations," *SIAM Review*, vol. 21 pp. 1-17, 1979.

[ShaGor 75]   L.F. Shampine and M.K. Gordon, *Computer Solution of Ordinary Differential Equations*. San Franscisco, CA: Freeman, 1975.

[ShaWat 77]   L.F. Shampine and H.A. Watts, "The Art of Writing a Runge-Kutta Code, Part I," in *Mathematical Software*, J.R. Rice (ed.). Orlando, FL: Academic Press, 1977, pp. 257-276.

[SreAga 92]   V. Sreeram and P. Agathoklis, "Design of Linear-Phase IIR Filters via Impulse-Response Gramians," *IEEE Trans. Signal Processing*, vol. 40, no. 2, pp. 389-394, 1992.

[Stasin 91]   R. Stasinski, "The Techniques of the Generalized Fast Fourier Transform Algorithm," *IEEE Trans. ASSP*, vol. 39, no. 5, pp. 1058-1069, 1991.

[Strass 69]   V. Strassen, "Gaussian Elimination is Not Optimal," *Numer. Math.*, vol. 13, pp. 354-356, 1969.
[Temper 83a] C. Temperton, "Self-Sorting Mixed-Radix Fast Fourier Transforms," *J. of Computational Physics*, vol. 52, no. 1, pp. 1-23, 1983.
[Temper 83b] C. Temperton, "A Note on Prime Factor FFT Algorithms," *J. of Computational Physics*, vol. 52, pp. 198-204, 1983.
[Temper 85]   C. Temperton, "Implementation of a Self-Sorting In-Place Prime Factor FFT Algorithm," *J. of Computational Physics*, vol. 58, pp. 283-299, 1985.
[Tustin 47]   A. Tustin, "A Method of Analyzing the Behavior of Linear Systems in Terms of Time Series," *J. of the IEE, Part IIA*, vol. 94, pp. 130-142, 1947.
[VanLoa 90]   C. Van Loan, "A Survey of Matrix Computations", Report CTC90TR26, Cornell Theory Center, Cornell University, Ithaca, NY, November 1990.
[VlcUnb 89]   M. Vlcek and R. Unbehauen, "Analytical Solutions for Design of IIR Equiripple Filters," *IEEE Trans. ASSP*, vol. 37, no. 10, pp. 1518-1531, 1989.
[Welch 67]   P.D. Welch, "The Use of the Fast Fourier Transform for the Estimation of Power Spectra: A Method Based on Time Averaging Over Short, Modified Periodograms," *IEEE Trans. AU*, vol. 15, no. 2, pp. 70-73, 1967.
[Winogr 78]   S. Winograd, "On Computing the Discrete Fourier Transform," *Mathematics of Computation*, vol. 32, no. 141, pp. 175-199, 1978.
[Zverev 67]   A.I. Zverev, *Handbook of Filter Synthesis*. New York: John Wiley and Sons, Inc., 1967.

CHAPTER 4

Preparing the Data Flow Graph for Partitioning

Partitioning the data flow graph into tasks and allocating them on the processors is based upon the structure of the graph and on information about the communication and execution costs.

In the preparatory steps, a data flow graph representation of the algorithm to be parallelized is first created. The graph's edges are then analysed for the volume of communications they carry. Subsequently, the graph is expanded in order to better exhibit the inherent parallelism. In the last step, each edge is assigned the communication costs it incurs, and each node is labelled with its execution costs. After these procedures, all the necessary data are available for statically partitioning and allocating the data flow graph.

4.1 Data Flow Graph Generation

Once the algorithm to be parallelized has been formulated in SISAL ([MGSkAl 85]), it is converted into a data flow graph by the standard compiler provided by the Lawrence Livermore National Laboratory (LLNL), which runs on a variety of host computers. The data flow graph is structured hierarchically and consists of edges with associated data types and nodes. Nodes representing basic operations such as addition and subtraction are called "simple" nodes. "Compound" nodes, e.g, Forall nodes, Iter nodes, and Select nodes (representing if-statements), are composed of several subgraphs. A Forall node consists of three subgraphs for range generation, for the body of the loop, and for returning the results. The graphs are described in textual form in a format called Intermediate Form 1 (IF1) ([SkeGla 85]).

The compiler carries out some standard optimizations such as common subexpression removal, constant folding, and loop invariant expression pullout (see [SkeWel 85]). Therefore, such optimizations are omitted in further processing.

Figure 4.1 shows a sample SISAL code for the multiplication of a vector x by a constant vector u. It consists of the function *controlalgorithm* as defined in Chapter 2.

```
         function controlalgorithm(x: array[real];
                 NumberOfInputs: integer;
                 returns    real)
      let
         v := array [1: 1.0, 2.0]
      in
         for j in 1,NumberOfInputs
            u := v[j] * dbl(x[j])
            returns value of tree sum u
         end for
      end let
      end function

      function dbl(a: real; returns real)
         2.0 * a
      end function

      function main(input: array[real];
         returns    real)
         controlalgorithm(input, 2)
      end function
```

Figure 4.1 Sample SISAL Code for a Scalar Product

Inside this function another function *dbl* is called which simply computes the double of a value.

The data flow graph representing these operations is used for illustrating the graph transformations and analyses described in the subsequent sections.

The algorithm to be parallelized is the function *controlalgorithm*. It receives the vector x as input parameter. The result is the product of the constant vector v and of x which is first multiplied by two using the function *dbl*. The function *main* is necessary only for analysis of the compilation and communication volumes analysis. The edges of the data flow graph are labelled in IF1 with their type, but not with the size of the composed structures (i.e., the number of elements in an array is not specified). Hence it is necessary to pass the size of the input vector to the called function as an additional parameter *NumberOfInputs*.

The data flow graphs of the functions defined in the example given in Figure 4.1 are shown in Figure 4.2. This has been drawn with a tool called IF1 Display ([MitMur 91]).

The hierarchical graph structure is clearly visible. The graph of the function *controlalgorithm* contains one Forall compound node. This node consists of three subgraphs. The first subgraph generates the set of indices J = {1, 2} used in the body subgraph to extract the elements of the vectors v and x to be multiplied in each invocation of the loop body. The elements of x are first passed to the function *dbl*. The

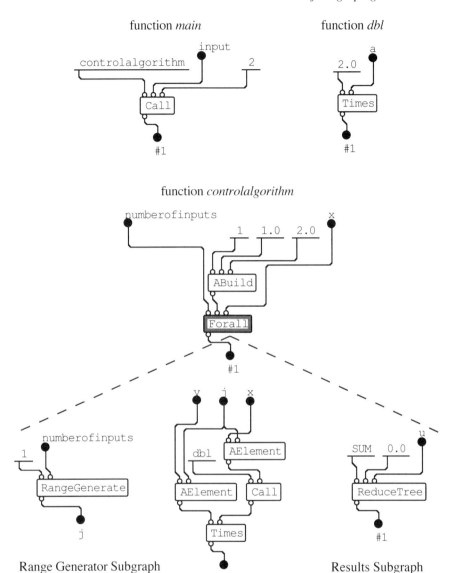

Figure 4.2 Data Flow Graph of the Scalar Product

results subgraph collects the results from each invocation of the loop body and computes the result of the node. In the example given above, all values are summed using a tree-shaped reduction scheme.

4.2 Communication Volume Analysis

As mentioned above, IF1 labels the edges only with the type of data, but not with its size. As far as basic data types are concerned (e.g., Boolean, integer, real), this does not pose any problem since it is known how much area is needed for storing data items of such types. But with the compound data type "array" it is impossible to know how many elements an array contains. Since the array sizes must be known for determining the communication cost, the whole graph has to be analysed with respect to the volume of communication.

This is done in a bottom-up, depth-first way. All edges of the function graphs are inspected for the size of their data, starting with the output edges. If the quantity of data is known because the data type is basic, then its size is annotated. If it proves impossible to find the size of the data type by inspection of the current edge, its source node is examined together with the input edges.

Depending on the node's function, useful information can be found (e.g., if the node was an ArrayBuild node, the number of array elements can be deduced from the number of input edges) and the amount of data transferred over the edge is attached to the edge. If the message size is still unknown, the search continues through all predecessor nodes of the current node. In the case where an Iteration node producing an array is encountered which has a termination criterion which is impossible to evaluate at compile time, a default value for the number of array elements is assumed. However, for signal processing applications this is a very unlikely case which should not occur.

The search is continued until either the size of the data of all edges visited is known or the edge leaves the top graph of the function *controlalgorithm*. In that case, the auxiliary parameter mentioned in Section 4.1 is interpreted as the number of elements in the input array. This is the only case where an array's size is impossible to determine, but the programmer has the necessary information and can provide it through this additional parameter.

4.3 Graph Expansion

Graph expansion aims at increasing the parallelism available in the data flow graph. The two main sources of parallelism are

- Forall Nodes
- Function Calls.

These two cases are discussed below in separate sections. Except for expanding those two nodes, some additional minor changes are made. These consist of removing NoOp (No Operation) nodes and two consecutive logical negation nodes. All other compound nodes (i.e., all forms of sequential loops expressed by "while" and "repeat...until") except Forall nodes are left untouched.

Structures building constant data (e.g., constant matrices) are duplicated so that each consumer of those values possesses a unique source of them. By this transformation, unnecessary data dependencies and data transmissions are eliminated, but at the expense of repeatedly generating constant values.

4.3.1 Forall Nodes

By definition, Forall nodes consist of a set of loop bodies which are independent of each other, e.g., in each body one element of a vector is processed. Due to this property, all bodies can be executed independently and concurrently. Hence the Forall nodes are replaced by their respective sets of body graphs with the inputs rewired accordingly. Indices are generated by special range-generation nodes and distributed to the appropriate sites. Since the results subgraph is eliminated as well, a new structure is built which collects the bodies' results and processes them according the reduction function (e.g., Sum, Product, Catenate).

Figure 4.3 shows the graph from Figure 4.2 with the expanded Forall node. Instead of the Forall node, the body subgraph is inserted twice, once for each invocation.

The range generator subgraph has been replaced by the RangeGenerateExp node which generates the indices for each body graph. Instead of the Results subgraph there

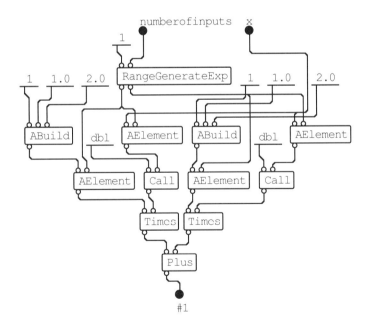

Figure 4.3 Graph with Expanded Forall Node (NumberOfInputs = 2)

58 Preparing the data flow graph for partitioning

is one single Plus node which represents the tree-summation reduction for two values.

Each body graph receives its own copy of the constant vector it needs for the scalar product. The ABuild (Array Build) node generating this vector has been duplicated in order to remove the dependency of both bodies from this node.

4.3.2 Function Calls

Function calls enhance a graph's parallelism because they increase its number of nodes. Their replacement by the function body has in view elimination of the function call overhead at run time. However, since this matters only for small functions, only a limited size (as far as the number of nodes is concerned) of the function graph is allowed for the replacement to take place. The size parameter is requested from the user. A positive effect of limiting function call replacement is that the growth of the graph is controllable by this size threshold.

Calls to intrinsic functions such as sine, cosine, and so forth are always left untouched.

In Figure 4.4, all calls to the function *dbl* have been replaced by the function's graph. The edges carrying the input parameters and the outputs have been connected directly to the respective ports of the function's nodes.

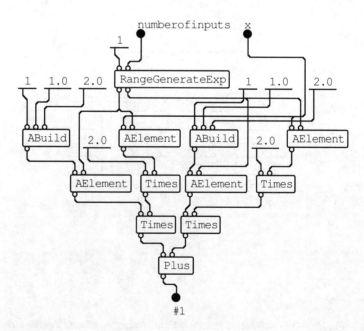

Figure 4.4 Graph with Expanded Forall Node and Replaced Function Calls

4.4 Execution Cost Analysis

For two reasons, it is necessary to know the execution costs of the data flow graph's nodes and of the data transmission over the graph's edges. First, the execution of the data flow graph is simulated during the partitioning phase in order to cluster the nodes into tasks so that the minimal execution time results. The time required for each operation must be known for this. Second, the weight of each task is relevant for balancing the load among the processors.

The data flow graph is traversed, and each edge and each node is assigned its execution cost. The edges' communication costs are determined according the model given below, and the nodes' cost is listed in Table B.4 shown in Appendix B.3.

4.4.1 Communication Costs

Since the communication network consists of bidirectional point-to-point serial links among the PEs, no unpredictable factors such as routing delays or an unknown number of intermediate stations appear in the communication costs. Nonetheless, the establishment of an accurate communication cost model remains a difficult task.

Although there is no difference in using links, whether they connect tasks located on one processor (called internal channels) or on two different processors (external

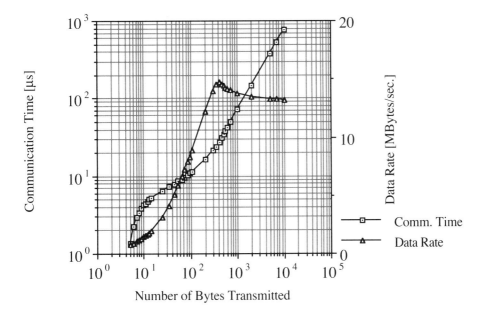

Figure 4.5 Communication Cost for Internal Channels

channels), the performance is different in each case. Figure 4.5 displays the communication time and the resulting net data rate for communication over internal links as a function of the length of the message to be transmitted. Figure 4.6 shows the communication time for external channels using the Transputer's (T800) bidirectional serial link at a speed of 20 MBit/second.

The graphs show actual measurements conducted in the laboratory. In both cases, communication time grows proportionally to the length of the message, as expected. However, the net data rate approaches its peak value only for message lengths of at least 200 bytes. As a consequence, messages should have this length in order to achieve efficient communication. However, in signal processing applications the graphs' edges seldom carry more than a few dozen values of 4 or 8 bytes each. For external communication, the theoretical peak rate of 1.8 MByte/sec is not reached, probably due to the two acknowledge bits sent for each byte received.

The ratio between the costs for internal and external communication is almost 1:10. This high cost for external communication justifies the efforts to reduce this kind of data transmission as much as possible.

The following mathematical model for the communication cost has been developed. All times are expressed in instruction cycles. At a clock frequency of 20 MHz each cycle

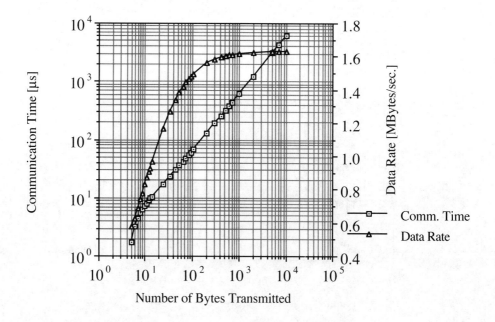

Figure 4.6 Communication Cost for External Channels

lasts 50 nanoseconds.

For internal communication, the communication cost can be computed according to the following formula:

$$c_{int} = T_{setup} + T_{memI} \times w \quad (4.1)$$

For external communication, additional terms appear in the formula:

$$c_{ext} = T_{setup} + T_{memE} \times w + T_{trans} \times b \quad (4.2)$$

The meanings and the values of the symbols are:

c_{int}	communication time for internal communication	
c_{ext}	communication time for external communication	
b	number of bytes transmitted	
w	number of long words (= 4 bytes) addressed	
T_{setup}	communication setup time	52 cycles
T_{memI}	memory access time, int. comm.	2 cycles/long word
T_{memE}	memory access time, ext. comm.	5 cycles/long word
T_{trans}	transmission time	11 cycles/byte

If the links are run at a speed of 10 MBit/sec, the transmission time T_{trans} doubles to 22 cycles/byte.

It is interesting to note that the memory access time depends on the number of long words fetched from memory rather than on the amount of bytes transferred. Therefore, this component of the cost remains the same whether one or four bytes are transmitted.

4.4.2 Node Execution Costs

Some of the graph's nodes represent basic operations such as addition and multiplication. For these, there is a simple equivalent in the processor's instruction set which takes a known amount of time to execute. However, there are a number of nodes for manipulating arrays (e.g., ArrayFill) with a varying amount of inputs and others that require more complex decisions, such as the Max or the Min nodes. The function of these nodes is emulated by a sequence of statements of the target PE's programming language. Determining their execution time accurately is difficult with figures from the PE's data book only.

Additionally, there are unexpanded compound nodes which contain whole subgraphs. These subgraphs are created by functions composed of the associated statement sequences of the nodes they contain. Finding the execution time is further hampered by the fact that for many of the nodes their exact number of iterations is unknown (e.g., for a While loop node).

Further variations of the execution costs are caused by the fact that most nodes accept a wide range of data types for their input and output values, whereas most programming languages are strongly typed so that additional type conversion routines have to be used.

For all those reasons, not only were theoretical studies undertaken but execution time measurements were also conducted for all major kinds of nodes and all possible data types. The results of those measurements have been assembled in a table showing the execution time for each node for every legal data type of the arguments (see Table B.4 in Appendix B.3).

4.5 References

[Jha 90] M. Jha, "Preliminary Results on some Parallel Linear Algebra Applications on Transputer Networks," School of Information Sciences, Hatfield Polytechnic, Numerical Optimization Centre, Technical Report No. 231, April 1990.

[MGSkAl 85] J. McGraw, S. Skedzielewski, S. Allan, *et al.*, "SISAL: Streams and Iteration in a Single Assignment Language, Language Reference Manual, Version 1.2," Lawrence Livermore National Laboratory Report LLL/M-146 Rev. 1, 1 March 1985.

[MitMur 91] S. Mitrovic and S. Murer, "A Tool to Display Hierarchical Acyclic Dataflow Graphs," in *Proceedings of the International Conference on Parallel Computing Technologies*, September 1991, Novosibirsk, USSR, pp. 304-315.

[Sarkar 89] V. Sarkar, *Partitioning and Scheduling Parallel Programs for Multiprocessors*, Research Monographs in Parallel and Distributed Computing. London; Cambridge, MA: The MIT Press, and London: Pitman Publishing, 1989.

[SkeGla 85] S. Skedzielewski and J. Glauert, "IF1–An Intermediate Form for Applicative Languages," Lawrence Livermore National Laboratory Report LLL/M–170, July 31, 1985.

[SkeWel 85] S.K. Skedzielewski and M.L. Welcome, "Data Flow Graph Optimization in IF1," in: *Functional Programming Languages and Computer Architecture*, Jean-Pierre Jouannaud (ed.), Lecture Notes in Computer Science vol. 201, Berlin a.o.: Springer-Verlag, 1985.

CHAPTER 5
Partitioning the Data Flow Graph into Tasks

5.1 Building Tasks

This chapter treats the problem of defining tasks by clustering nodes of the data flow graph. Tasks consist of at least one node and are executed as an entity.

The tasks are distributed to the processing elements (PEs) of the system in the allocation step. The execution of the nodes of a task is serialized even if some nodes could be executed in parallel. Each PE can host several tasks. Then tasks are sometimes called processes, meaning "a unit of activity characterized by a single sequential thread of execution, a current state, and an associated set of system resources" ([IEEE 88]).

Once a graph representation of the computations to be performed has been established, tasks can be built by clustering the nodes. This need not necessarily be done since the nodes could be distributed directly to the processing elements (PEs). However, the aims of partitioning a data flow graph into tasks are as shown in Table 5.1.

Table 5.1 *Aims of Partitioning*

1. preserve the parallelism inherent to the problem
2. minimize the task administration cost at run-time
3. minimize the total communication between the tasks
4. minimize the communication and synchronization overhead

Not partitioning the graph into tasks is equivalent to creating one task per node. This clearly exhibits all the parallelism in the problem. However, clustering several nodes in a task does not necessarily decrease the amount of work potentially executable in parallel. Since each data flow graph represents a partial ordering of the nodes, there always exist some nodes which cannot be executed in parallel but have to be computed sequentially. When such nodes are placed in one task, no parallelism is lost.

As an example, imagine computing a = 2 × log b. Assuming there is a node performing the "log" operation, the node computing the multiplication depends only on the output of the "log" node and the constant input "2". This means that the result "a" comes from a path containing two nodes with just one external input "b" at the origin. These two nodes can be placed in one task without losing any parallelism because the second node is data dependent on the first. However, it is important to maintain as much parallelism as possible to keep many processing elements busy, because only then can the total execution time be kept to a minimum. Partitioning is also an issue on single-processor systems in order to obtain fragments of a program which can be handled efficiently by the scheduler ([Paige 77]).

Even if the tasks contain only two nodes, the number of tasks is already reduced by half. As a consequence, at run-time only half the number of tasks have to be administered. This greatly reduces the overhead cost of creating, starting, and stopping the tasks at run-time. Having just one task on each PE could be considered optimal. However, this is not the case. Since generally the number of PEs is much smaller than that of nodes in the graph, each task would have to contain many nodes in order to have only one task per PE.

Placing one large task on each PE would enforce some serialization of the computations, i.e., the introduction of a total ordering of the nodes instead of the partial order defined in the original graph. If the computations are totally ordered, one specific architectural feature present in many types of PEs can no longer be exploited. It is now not possible to overlap computations of one task with the communications of another.

However, this property helps to reduce the total execution time considerably, especially in the case of fine-grain parallelism where the execution times of the nodes are of the same order of magnitude as the communication cost. Therefore, it is desirable to have more than one task per PE in order to allow overlapping communication and computation, but not too many in order to avoid high scheduling cost ([Cvetan 87]).

Communication is a large part of the total execution cost of a program, as outlined in Chapter 4. It is therefore important to keep overall communication among the tasks low. Every connection between two nodes which lies within one task is produced by a variable and is free. Thus, if nodes can be clustered in such a way that many edges come to lie within tasks, the total execution cost is lowered.

Due to the data flow principle special procedures for synchronization such as signals, semaphores, monitors, etc. are not necessary. However, apart from pure transmission cost, data communication between the processes causes further expense. The communication cost model defined in Chapter 4 shows the high overhead for set-up. The ratio between set-up and transmission cost can be lowered only by increasing the message length.

Unfortunately, nothing can be done about this when partitioning a data flow graph. Even worse, if several edges of the graph have to be mapped on one physical connection because too few links are available, additional time multiplexing must be used for the links. This causes additional communication overhead. The only way to keep this cost low is to have as few edges connecting the tasks as possible. However, there is often little choice in this since the interconnections are given by the structure of the problem.

In sum, the data flow graph must be partitioned such that:

- Sufficient parallelism remains among the tasks (aim one)
- The number of tasks per PE is reduced to an optimum (aim two)
- Heavily interconnected nodes are placed in the same task (aim three)
- Tasks have few interconnections among them (aim four)

Above all, the total execution time of the data flow graph should become minimal. Aims three and four are in conflict with aim one, because minimizing communication very quickly means eliminating parallelism. Furthermore, it is not clear what are the optimal values for the task size and the number of tasks. Therefore it is necessary to formulate some heuristic measure for the quality of a partition. This is proposed in Section 5.4, together with a partitioning algorithm.

5.2 Approaches Described in the Literature

One approach taken is purely graph theoretical. The method is to find the components of a graph so that the "min-cut" problem is solved, i.e., groups of nodes are identified so that the number of edges cut by the task-separation lines is minimal ([Barnes 82]). A variant of the problem is to find the n-connected components of the graph, that is, the components which are connected only at n points. However, really efficient algorithms exist only for n = 2 and n = 3 ([HopTar 73], [Even 75], [Even 79]).

Even if algorithms terminating in polynomial time do exist, minimizing the communication overhead is difficult since this approach considers only structural properties of graphs and does not take into account the reduction of communication cost when nodes are merged in tasks. (Polynomial time means that the number of computations and therefore the time required for performing them is expressable by some polynomial formula containing the problem's size.) Above all, real program graphs rarely have such regular structures that they can be decomposed into enough n-connected components.

Taking into account communication and scheduling costs makes matters even worse. Communication costs for an edge are different whether the edge crosses a task boundary (and possibly a processor boundary) or not. The cost of scheduling the tasks varies considerably with the number of tasks. Depending on the scheduling strategy, the cost is a nonlinear function of the task number. Since each partitioning strategy aims at minimizing the total execution costs (communication, run-time scheduling, and computation), and as these costs are very dependent on the particular partition of the computations, it is understandable that partitioning is a difficult problem to solve.

Sarkar proves in [Sarkar 89] that finding an optimum partition of a data flow graph with minimum total execution time is NP-hard in the strong sense. NP-hard denotes solving a problem that is NP-complete, meaning that computing the solution is not complete after polynomial time. That is, solving a problem with N input values requires exp(2N) computations, e.g., instead of a number expressed as a polynomial of N, such as $N^4 + 3N^6$. Hence only heuristic approaches are tractable, because finding the best solution takes too much effort to be feasible. With a heuristic method probably only suboptimum solutions will be found, but in a much shorter time which may still be considerable.

All papers dealing with scheduling computations address the topic of partitioning the application into pieces of an appropriate size. However, most authors assume that this has already been done previously. In many of the remaining cases, partitioning is only outlined. In [KoMeVR 88], "maximizing locality of memory reference" is mentioned as an aim. Identifying "maximal connected subgraphs" is given as a goal in [HoPaFe 86].

Huang ([Huang 85]) develops a formal model of software partitioning which relies on the module precedence relations for generating tasks from modules specified by the programmer. The model observes the constraints imposed by PE throughput, memory space available, and maximal allowed task execution time. The software partitioning efficiency σ is defined as

$$\sigma = \frac{\sum_{k=1}^{N} E(T_k)}{\sum_{k=1}^{N} [E(T_k) + O(T_k)]}$$

and has to be maximized under various constraints. $E(T_k)$ stands for the execution cost of task T_k and $O(T_k)$ for the execution overhead, including the task scheduling and the communication costs. It is shown that maximizing the partitioning efficiency also minimizes total execution time. When σ is maximal, the overhead is minimal and thus the execution time is also minimal.

The crucial point is how to choose the maximal allowed task execution times, and what strategy to employ to find the maximal partitioning efficiency. Both points are stated to be currently unsolved. However, the concept of partitioning efficiency can be a valuable tool to measure the quality of a partition obtained with any method.

In [KoMePe 88], partitioning is approached in two steps. In order to reduce the number of nodes of the data flow graph, so-called "supernodes" are formed "by gathering nodes directly connected by arcs in the data flow graph". This new graph is called CDFG (Compressed Data Flow Graph). Subsequently, the CDFG is divided arbitrarily among an unbounded number of PEs. This initial partition is iteratively improved by moving single nodes from one partition to the neighbouring one until the restrictions concerning the number of interconnections among the PEs are satisfied. By this method a modified min-cut problem is solved.

Campbell ([Campbe 85]) limits the complexity of partitioning by a divide and conquer approach. He creates a so-called module graph using the parse tree. Each module corresponds to one segment of the program as specified by the programmer, i.e., modules represent a function, a body of a loop, etc. Each module is then partitioned using a topological sort by Breadth-First Search (BFS) and its transitive closure. Partitioning is done so that a cost function becomes minimal. Two cost functions are suggested: one measuring communication cost, which is proportional to the distance a message has to travel, the other measuring parallel processing cost, which increases if two nodes executable in parallel are placed on the same PE.

This modularization concept offers the great advantage of keeping small the size of each graph to be partitioned. Computations such as the determination of the graph's transitive closure can then be performed quickly, even though they require a computational effort of $O(N^3)$, when N nodes are present in the graph. However, the amount of parallelism contained in the modules depends on the formulation the programmer chose for the computations since it is derived from the parse tree representing the source code's structure. Depending on the programmer's skill and the capabilities of the programming language, severe performance degradation could result. But since the input language used is HDFL (Hughes Data Flow Language), a derivative of the functional language VAL ([AckDen 79]), the negative effects of this approach are limited.

The last two approaches described are employed by Gaudiot and Lee, and by Sarkar. They all use SISAL ([MGSkAl 85]) as an input language and work with data flow graphs described in the Intermediate Form 1 (IF1) ([SkeGla 85]). As mentioned in Chapter 2, the work presented here uses the same representation. For this reason, these projects are examined in more detail.

Gaudiot and Lee ([GauLee 88]) aim at exposing maximum parallelism and achieving minimum communication cost. However, they do not state how this is can be done. Furthermore, they do not consider the execution cost of the nodes. Therefore, their partition will hardly yield a low total execution time.

From the original IF1 graph, a Partitioned Data Flow Graph (PDFG) is created using a Program Structure Graph (PSG). In this PDFG so-called block nodes are introduced which represent sub-graphs of compound nodes. For partitioning, the hierarchical graph is traversed in a Depth-First Search (DFS). Partitions are formed from each simple node and from all the simple nodes attached to a block node. Generally, the resulting tasks are small.

Sarkar proposes two heuristic approaches for partitioning. One is designed for systems with run-time scheduling cost; he calls this the macro-data flow approach. The other proposal, called the compile-time scheduling approach, is intended for statically assigning work to an unbounded number of processors in the so-called internalization pre-pass. These methods, given in [Sarkar 89], are improved versions of results presented in [SarHen86].

In the macro-data flow approach, an initial partitioning of the hierarchical data flow graph is set up by expanding all nodes with an execution time greater than T_{min}. Through this parameter the granularity of the initial partition is controllable. Tasks are then successively merged until only one remains. For each partition Π_i the cost $F(\Pi_i)$ is computed. Finally, the partition yielding minimal cost is chosen. The cost $F(\Pi_i)$ is defined as max(critical path length, sum of all scheduling overhead cost). In each step the pair of tasks is merged, which reduces the overhead cost most and results in the least increase in the critical path's length. If the graph has N nodes, the maximum complexity of the partitioning algorithm is given as $O(N^3)$, while experiments indicated a complexity in the region of $O(N^{1.2})$.

When doing compile-time scheduling, the graph is first expanded. Function calls are replaced by the functions and compound nodes are expanded (mainly loop unrolling). Again the execution time threshold T_{min} is used, and a node must be a bottleneck node in order to be chosen for expansion. A node is said to be a bottleneck if all the simple nodes executable in parallel to it do not contain sufficient work to keep busy all the processors in the system. This node expansion is done in linear time, i.e., the time required for performing the computations grows linearly with the size of the problem.

The internalization pre-pass actually belongs to the scheduling of the graph on the processors, and only the number of PEs is unbounded. Each PE hosts only one task. The communication cost F_c of the edges is taken into account. A processor assignment PA^∞ is sought which yields the minimum parallel execution time. It is proved that the solution of this problem is also NP-complete in the strong sense. Therefore, a heuristic straightforward solution is proposed. Backtracking methods are excluded due to exponential execution times.

The algorithm works as follows. Initially, all nodes are allocated a task of their own. All edges are examined in descending order of their communication cost F_c. The two nodes connected by the edge are joined in the same task (i.e., on the same PE) if merging them does not increase the parallel execution time of the graph. Joining the two nodes means replacing the external communication over the edge by internal communication, hence the name of the procedure. The complexity of the internalization is maximally $O(N^2)$, depending on the structure of the graph, with practical values rather in the range of $O(N)$.

Simulations by Sarkar indicate that compile-time scheduling offers better speed-up than the macro-data flow approach due to less run-time overhead. However, the assumption of only one serial task on each PE means that more parallelism than is necessary is sacrificed when the actual hardware offers a scheduler with minimum overhead. The macro-data flow approach has been partially incorporated into the research version POSC (Partitioning and Optimizing SISAL Compiler) ([SarCan 90]) of the SISAL compiler distributed by the Lawrence Livermore National Laboratory.

5.3 Deadlock Avoidance

As soon as the execution of computations no longer has total ordering, measures have to be taken to prevent any two instructions in a program waiting for each other's output to start executing. Two reasons can cause a program to lock: either the data flow graph's structure, or the implementation of the mechanisms for executing a data flow graph which makes decisions for serializing the execution.

Structural locking is detectable from the data flow graph's transitive closure, although it is expensive to compute. It should never occur when transforming real-world

Deadlock avoidance 69

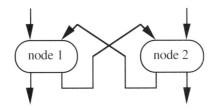

Figure 5.1 Two Nodes That Will Never Fire

computations into a data flow graph. Otherwise possibly nonterminating computations are represented where some equations are connected so that a never-ending recursion is generated. A simple example of a combination of two nodes which will never be able to fire is shown in Figure 5.1. It is assumed that no data tokens are present initially. Each of the nodes can only fire if it receives a data token at its external input and at the input originating at the output of the other node. But since the other node is unable to fire until it receives the output token from the first node, each node cannot proceed until the other has produced a result. This represents a typical deadlock situation where interconnected cycles exist in the data flow graph.

A data flow graph can be translated into a Petri Net. For this kind of graph there exist formal methods for analysing the possibility of deadlock (see [Starke 90]). An example of locking caused by the implementation is shown in Figure 5.2.

On true data flow computers, the the data tokens' order of arrival is irrelevant, since they are stored in buffers. During execution, it is periodically checked whether all the values needed for a node are present.

When implementing a data flow graph on a machine with conventional architecture, i.e., with sequential execution of code segments, problems can arise when the communication statements are incorrectly ordered. Sequential execution can be avoided by creating parallel processes for all communication statements. This corresponds to implementing true nonblocking communication. With nonblocking communication the sender of a message can continue processing as soon as the message has been written to a buffer. There it is eventually read from by the receiver at a later time. Blocking communication stops the sender until the receiver is ready to execute the "read" statement and vice versa. Through this mechanism communication also serves as a means of synchronization.

However, this causes too high an overhead to be feasible in most cases where the execution time is critical. Therefore, if blocking communication is the default message-passing mechanism, input and output from the processes have to be handled carefully, as explained below.

When two parallel edges connect two nodes, as shown in Figure 5.2 a), the execution is evidently blocked if the sequence of the output statements of node 1 is different from the sequence of the input statements of node 2. Fortunately, such a

70 *Partitioning the data flow graph into tasks*

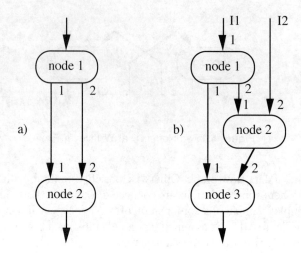

Figure 5.2 Example of Implementation Locking

configuration does not occur in IF1 graphs. However, if two tasks each containing several nodes are connected in this way, the same problem arises. The only way to exclude locking in this case is to make sure that edges leave the task only from the last node. Thus the possibility of parallel edges is eliminated.

For the configuration with three nodes given in Figure 5.2 b), the following two orderings of communication statement execution are assumed.

Ordering 1

	first	second
node 1	output 2	output 1
node 3	input 2 (1)	input 1 (2)

Ordering 2

	first	second
node 1	output 1	output 2
	input 2	input 1

The order of all operations not mentioned is irrelevant. With Ordering 1, everything works smoothly. As soon as node 1 receives the input data item, it outputs the result at output 2, where it is received by the waiting node 2. As soon as node 2 has the other input data item, the output is activated and node 3 receives the result at input 2. Then node 1 can send the token from output 1 to input 1 of node 3. If the sequence of the input statements of node 3 is reversed as indicated in parentheses, node 1 can proceed directly without having to wait until node 3 has finished receiving the result from node 2. Both orders of the input statements of node 3 will work.

Consider Ordering 2. Node 1 receives the input item and proceeds to forward its result through output 1. However, since node 3 awaits the first data token at input 2, node 1 is unable to deliver its data token at output 1 and cannot send a value through output 2 to node 2. Therefore the computations are blocked and no result can be generated. The computations will terminate only if the sequence of the input statements of node 3 is reversed.

Basically, the problem is due to the fact that there are two paths from node 1 to node 3. One connects the two nodes directly, representing a direct dependency. The other path has intermediate nodes, representing an indirect dependency. If the direct dependency edge is provided with the output data first, then the order of the input statements at the end node is critical, as shown with Ordering 2 in the above example. If the indirect dependency edge is first activated, any ordering of the end node's inputs will allow the computations to terminate. The execution time may, of course, vary because it is possible to introduce unnecessary delays.

Since the expense of finding all such paths in a graph is prohibitive, an equivalent approach is taken. It is similar to the method called Length of Longest Output Path (LLOP) in [HoIra 83] which is used for the decision of which of two enabled nodes to execute first. Each output is annotated with the longest path to the result node of the graph, and the cost of executing it. The length of a path indicates how many nodes are traversed along the path, and the cost is the sum of the execution cost of all these nodes.

The graph is traversed in a bottom-up fashion in a modified breadth-first search (BFS). For each node in the graph an order of its output ports is then defined. The sort criteria are listed in Table 5.2. A second criterion is introduced since it is quite possible that the data flow graph consists of some geometrically identical subgraphs due to a regular structure of the computations (e.g., in a matrix multiplication). Several paths with the same length exist in that case. If the order of the output statements introduced by the rules just introduced is respected, no execution order of the input statements of the nodes in the graph can cause the graph to deadlock.

Table 5.2 *Criteria for Determining the Execution Order of a Node's Output Statements*

Primary Sort Criterion	Longest Path to Result Node First
Secondary Sort Criterion	Longest Execution Time Along Path First

Apart from guaranteeing the freedom from deadlock in a graph's execution, this ordering reflects the idea of evaluating the nodes of the critical path first in order to obtain a minimum execution time. While the longest path from a node to the result node need not necessarily be the critical path, the output port having the longest path to the result node is likely to be found on it.

5.4 Partitioning the Data Flow Graph

As outlined in Section 5.1, several conflicting aims exist when a data flow graph is partitioned. Since the total execution time is the most critical issue for signal processing and control applications, partitioning is formulated as a minimization problem with respect to program execution time. However, preserving sufficient parallelism among the tasks at that stage of processing is equally important, because the number of processing elements (PEs) necessary to perform the computations is still open.

5.4.1 Rules For Building Tasks

In the previous steps, the data flow graph has been assigned all the necessary information about communication cost and execution cost. Now the tasks are formed according to the rules defined in Figure 5.5 below. The formalism for building the tasks is outlined in more detail in Appendix B.4. The most important processing steps are described in the following sections.

In a first step, the graph is prepared for partitioning. This includes computing the adjacency matrix and the transitive closure of the graph and defining one task for each node. Each task is assigned a PE of its own. This is the configuration exhibiting maximum parallelism as well as maximum communication cost and the lowest PE utilization.

Subsequently, the execution time is minimized by clustering the tasks which so far contain only one node. The minimization runs in a loop which is executed until the minimum execution time is found. In each invocation of the loop, all linked pairs of tasks are merged temporarily if certain criteria (see below) are fulfilled. For this new task configuration, the completion time is determined by simulating the flow of the data tokens through the graph.

If the time is lower than the minimum so far encountered, the edge linking the investigated task to its successor is recorded together with the new minimum execution time. Then the changes in the tasks are reversed and the next pair of tasks is examined. When all possible candidates for merging have been inspected, the task pair yielding the lowest execution time is merged in a new task, thus eliminating one external communication channel (an edge crossing the task boundary). This channel is replaced by communication through a variable. Then the next pair of tasks is sought.

When the minimum execution time has been reached, clustering continues for a while without affecting the execution time. The execution time is basically determined by the critical path which is no longer modified after a certain degree of processing. However, the other tasks grow in size. With larger and therefore fewer tasks the subsequent allocation step is accelerated.

Candidates for merging are found by inspecting the last node in a task. The criteria for merging the task of that node and the task of one of its successors are displayed in Table 5.3.

Partitioning the data flow graph 73

In Condition 1 only tasks of directly connected nodes are merged. Merging tasks of unconnected nodes does not make sense since no external communication channels are thus eliminated.

Table 5.3 *Rules for Merging Tasks*

1. The node must have a direct successor
2. The node may have only one external output edge
3. The node at the end of the output edge may only have external or constant inputs

Condition 2 ensures that the resulting new task does not have an edge leaving from intermediate nodes. This would violate the rule that the results may leave a task only through its last node. Figure 5.3 a) shows the situation where Condition 2 is violated. If the existing tasks *T1* and *T2* were merged in the new task *T1'*, the second (external) output of the last node of *T1* would leave the new task at a different location from the end. For this reason the last node in a task to be merged must not have two output edges.

By Condition 3 it is enforced that no parallelism is destroyed. Consider the situation

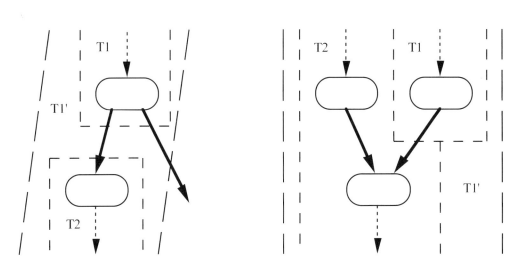

a) Two External Outputs   b) Successor Has Another Internal Input

Figure 5.3 Node Combinations Not Allowed for Combination

74 Partitioning the data flow graph into tasks

depicted in Figure 5.3 b). There, the two nodes from tasks *T1* and *T2* drawn side by side are executable in parallel since they are not dependent on each other. However, if tasks *T1* and *T2* were merged in the new task *T1'*, the execution of the two nodes would have to be serialized, thus destroying parallelism. The independence of the two nodes can be tested by inspecting the successor of the last node of task *T1*. If there is an input edge leading to the successor from another node in the same task as the successor node, then parallelism is present and the tasks may not be merged.

Figure 5.4 shows the correct situation for tasks *T1* and *T2* to be joined in task *T1'*. The last node of task *T1* possesses only one output to the first node in task *T2*, thus satisfying all three conditions defined in Table 5.3.

The tasks constructed by these rules consist of chains of nodes. Of each node's input channels, several may be external but only one internal. Only one output edge is allowed, which leads to a node belonging to the same task. Only the last node in the chain is allowed to have edges leaving the task. In Figure 5.5, the algorithm for building the tasks of a graph *fgr* is shown.

The data flow graph partitioned in that way into tasks still possesses the same amount of parallelism as though each node were treated as a task of its own. As a consequence, the resulting tasks are small, mostly containing two to three nodes.

5.4.2 Preparation of the Data Flow Graph

The preparation of the graph includes several steps briefly outlined here. First, the adjacency matrix is constructed. The original IF1 graph provided by the SISAL compiler is such that the adjacency matrix is strictly upper triangular. This property is preserved

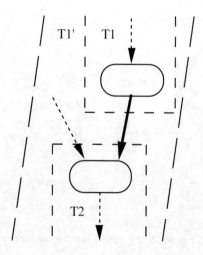

Figure 5.4 Node Configuration for Combination of Tasks

```
PARTITION (fgr, minfactor):

prepare graph
sort_outputs(fgr)
mintime := simulate(fgr)
repeat
    for all nodes nd in fgr with only one external output oed do
        if sink node of oed has only external and constant inputs then
            join sink node of oed and nd in one process
            replace oed by variable
            sort_outputs(fgr)
            executiontime := simulate(fgr)
            if executiontime<mintime then
                minedge := oed
                mintime := executiontime
            end if
            separate the two nodes again, restore original state
        end if
    end for all
    join nodes connected by minedge in one process
    replace minedge by variable
until execution time has reached the minimum
end partition
```

Figure 5.5 Partition of a Data Flow Graph

throughout all transformations of the graph. Therefore, computation of the transitive closure is performed with a modified version of Warshall's algorithm, saving much work (complexity $O(N^3/6)$ instead of $O(N^3)$ with N nodes in the graph).

For each node of the graph, a process is created and located on a PE of its own. During the whole partitioning process, each task is hosted by a separate PE. Only after allocating the processes on the PEs it is possible for several tasks to share the same PE.

Nodes which have constant inputs only and therefore no predecessors are treated in a special way. Nodes with constant inputs only are immediately joined to the tasks of their successor, if there is only one. It does not make sense to preserve parallelism in structures which generate constant data and which are evaluated before run-time.

Ordering the outputs has been explained in Section 5.3 above. It is done prior to the first simulation of the graph's execution and subsequently after each modification of the task structure.

5.4.3 Simulation of the Data Flow Graph's Execution

Each time two tasks are merged temporarily during the search for the lowest execution cost, the execution of the data flow graph is simulated in order to get an estimate of the time needed to perform the computations. Another purpose of the simulation is to establish a partial order of the nodes' input operations. This order is not mandatory to ensure that the execution does not block, but it does have a positive impact on the execution time of the computations. It represents the order of arrival of the input data with respect to time.

During partitioning, the data flow graph consists of a set of tasks or processes, each containing one or more (connected) nodes. Each process is located on a PE of its own. Edges between nodes in the same task represent communication by a variable, those between nodes in different tasks denoting communication over inter-processor links. Inter-task communication is costlier than communication by variable, since the transmission speed is lower and additional overhead for set-up on both the sender and the receiver results.

On the multiprocessor assumed to consist of T800 Transputers, communication is blocking, i.e., transmission starts only when the transmitter and the receiver have reached the corresponding points in their respective programs. They continue the execution of the program only when all data have been transferred.

For the simulation of the graph's execution, each PE has a hypothetical clock which is used to stamp the input and output events. It is driven by the availability of input data and the operations performed on the PE, i.e., by the time taken by computations and input/output operations. Time spent waiting for input or output is also taken into account. Therefore, when the final result has been generated, the clock of the PE on which the last node of the data flow graph is located shows the time taken by the execution of the whole graph. Another counter on each PE sums the times consumed by every action of the PE. From this last counter the utilization of the PE can be computed.

The simulation process is portrayed in detail in Appendix B.4. A short description is given below to explain the basic principle.

The simulation starts by inserting all nodes which have received their input data (i.e., the input node, nodes with only constant inputs) into a circular list. This list is processed until it is empty.

For each node in the list, the output edges are inspected in the order previously established for avoiding deadlocks. For each output edge, the set-up time is added to the PE clock. This is called output time. It is the earliest time that data are ready for transmission. Then it is checked to see whether its successor node has input data available at each of its inputs. If the answer is no, the next node in the list is inspected. In the affirmative case, the successor node is processed (see below), and the clock is advanced to the "rendezvous time", i.e., to the time the data transmission has been completed. Then the next output is inspected. When all outputs of a node have been processed, it is removed from the list.

Analysis of the output edge of a node in the list may have been postponed after setting up communication due to the fact that the successor node was not ready. Then two possibilities exist. Either the situation has not changed, in which case the successor node is inspected and processed if possible, as described above, or the successor has meanwhile become ready for execution and has been processed. The output edge thus already carries a rendezvous time stamp, and the PE's clock is advanced to that time without further processing.

When a successor node is processed, all outputs connected to it are ready to transmit data, marked with the output time indicating at what time data are available. The inputs are sorted according to the time that data are available and are inspected in that order. If the node is the first one in the task, the input is marked with the input time, i.e., with the PE clock plus the communication set-up time. The rendezvous time is then the greater of either the input time or the output time of the corresponding output plus the data transmission time. The clock is set to the rendezvous time, since processing can then continue.

If the node has a predecessor in the same task, it may happen that data from the predecessor arrive much later than data at an external input of the node under consideration. It will then save time if the external input is served before the function of the predecessor node is performed. If this situation occurs, the external edge and its predecessor node are marked accordingly.

Normally, the function of a node is not "executed" (its execution time added to the clock) immediately. The decision when to execute is postponed to the time when its successor in the same task is inspected. But if there is no successor, the execution time is added to the clock and to the counter, and the first output is inspected. Then the recursive analysis of its successors starts. The node is inserted into the circular list for further inspection of its outputs.

After the simulation, the clock of the PE on which the edge carrying the final result of the graph is located shows the total execution time.

5.4.4 Computational Complexity Analysis of Partitioning

It is difficult to determine the exact amount of work done during partitioning. The reason is that an iterative search for the task configuration causing minimum execution time is conducted. However, it is possible to estimate the order of magnitude of the number of operations performed when taking into account results obtained in experiments. Inserting a node into a list or setting the time values for an edge is understood by "one operation". The number of nodes in the graph is indicated by N. The complexity of the processing steps of partitioning is shown in Table 5.4. The most important information is how many times the minimization loop is executed. The experiments conducted so far have shown that, after partitioning, around N/2 tasks exist. Since the initial number of N tasks is reduced by one with each iteration, the loop is executed approximately N/2 times.

The complexity is in the magnitude of $O(N^4 + N^3)$, which is relatively high. However, since this search is done off-line and the gain of information is considerable, this effort is justified.

Table 5.4 *Computational Complexity of Partitioning*

Function Performed	Operation Count	Total
Preparatory Steps:		
- build a list of the nodes	N	
- compute the transitive closure	$N^3/6 - N^2/2 - 2N/3$	
- define the PEs	N	
- order the outputs of nodes	$2N$	
- simulate the graph's execution	$2N^2$	
Total Preparation:	$\dfrac{N^3}{6} - \dfrac{5N^2}{2} + \dfrac{10N}{3}$	$\dfrac{N^3}{6} - \dfrac{5N^2}{2} + \dfrac{10N}{3}$
Minimization Loop:	$\sim N/2$ iterations	
- order the outputs of nodes	$2N$	
- simulate the graph's execution	$2N^2$	
Total per Iteration:	$2N^2(N+1)$	
Total Minimization Loop:		$\sim N^3(N+1)$
Grand Total (approx.):		$N^4 + \dfrac{5N^3}{6} - \dfrac{5N^2}{2} + \dfrac{10N}{3}$

5.4.5 Summary of the Properties of the Partitioning Technique

The partitioning procedure described in this chapter has been developed specifically for signal processing and control algorithms. Due to the simple properties of these algorithms, recursive computations are not treated in a special way. Therefore, iterative computations are not implemented in periodic schedules which might lower their execution time. Furthermore, recursive computations which in fact are very untypical in signal processing must be reformulated as iterative computations before the partitioning process. This is also required by the aim to exhibit and preserve parallelism in the computations. The average number of nodes per task is relatively small as a consequence.

Since the computations are to be carried out in real-time the main emphasis is put on minimizing the execution time. This is verified by repeated simulation of the execution of the data flow graph. Special care is taken to reduce the communication cost by eliminating as many external and internal channels as possible and replacing channels by variables. These decisions are possible since it is assumed that the tasks are allocated statically to the processors.

Also of importance for real-time computations is the fact that the tasks are formed in a way so that deadlocks are prevented. Such conflicts cannot be resolved by the simple round-robin scheduler running on each processor.

5.5 References

[AckDen 79] W.B. Ackerman and J.B. Dennis, "VAL–A Value-Oriented Algorithmic Language: Preliminary Reference Manual," MIT Laboratory for Computer Science, Technical Report MIT/LCS/TR-218, June 1979.

[Barnes 82] E.R. Barnes, "An Algorithm for Partitioning the Nodes of a Graph," *SIAM J. Alg. Disc. Meth.*, vol. 3, no. 4, pp. 541-550, 1982.

[Campbe 85] M.L. Campbell, "Static Allocation for a Data Flow Multiprocessor," in *Proc. Int. Conf. Parallel Proc.*, Aug. 1985, pp. 511-517.

[Cvetan 87] Z. Cvetanovic, "The Effects of Problem Partitioning, Allocation, and Granularity on the Performance of Multiple-Processor Systems," *IEEE Trans. Computers*, vol. 36, no. 4, pp. 421-432.

[Even 75] S. Even, "Algorithm for Determining Whether the Connectivity of a Graph is at Least k," *SIAM J. Comput.*, vol. 4, no. 3, pp. 393-396, 1975.

[Even 79] S. Even, *Graph Algorithms*, Computer Software Engineering Series. Potomac, MD: Computer Science Press, 1979.

[GauLee 88] J.-L. Gaudiot and L.-T. Lee, "OCCAMFLOW: A Methodology for Programming Multiprocessor Systems," *J. Parallel and Distr. Comput.*, vol. 7, no. 1, pp. 96-124, 1988.

[HoIra 83] L.Y. Ho and K.B. Irani, "An Algorithm for Processor Allocation in a Dataflow Multiprocessing Environment," in *Proc. Int. Conf. Parallel Proc.*, Aug. 1983, pp. 338-340.

[HoPaFe 86] Y.-C. Hong, T.H. Payne and L.B.O. Ferguson, "Graph Allocation in Static Dataflow Systems," *Computer Architecture News*, vol. 14, no. 2, pp. 55-64, 1986.

[HopTar 73] J. Hopcroft and R. Tarjan, "Dividing a Graph Into Triconnected Components," *SIAM J. Comput.*, vol. 2, no. 3, pp. 135-158, 1973.

[Huang 85] J.P. Huang, "Modeling of Software Partition for Distributed Real-Time Applications," *IEEE Trans. Softw. Engineering*, vol. 11, no. 10, pp. 1113-1126, 1985. Also reprinted in [ShaWan 89].

[IEEE 88] *Standard Dictionary of Electrical and Electronics Terms*, Fourth Edition. ANSI/IEEE Std. 100-1988. New York: The Institute of Electrical and Electronics Engineers, Inc., 1988.

[KoMePe 88] I. Koren, B. Mendelson, I. Peled, *et al.*, "A Data-Driven VLSI Array for Arbitrary Algorithms,", *Computer*, vol. 21, no. 10, pp. 30- 43, 1988.

[KoMeVr 88] C. Koelbel, P. Mehrotra and J. Van Rosendale, "Semi-Automatic Process Partitioning for Parallel Computation," NASA Contractor Report 18163, ICASE Report No. 88-16, NASA Langley Research Center, Hampton, VA, February 1988.

[MGSkAl 85] J. McGraw, S. Skedzielewski, S. Allan, *et al.*, "SISAL: Streams and Iteration in a Single Assignment Language, Reference Manual," Version 1.2, Lawrence Livermore National Laboratory Report LLL/M-146 Rev. 1, 1 March 1985.

[Paige 77] M.R. Paige, "On Partitioning Program Graphs," *IEEE Trans. Softw. Engineering*, vol. 3, no. 6, pp. 386-393, 1977.

[SarCan 90] V. Sarkar and D.C. Cann, "POSC–A Partitioning and Optimizing Sisal Compiler," in *Proc. Conf. on Supercomputing*, Amsterdam, 11-15 June 1990, pp. 148-163. Also: LLNL preprint UCRL-102737 Rev. 1, April 1990.

[SarHen 86] V. Sarkar and J. Hennessy, "Compile-Time Partitioning and Scheduling of Parallel Programs," *SIGPLAN Notices*, vol. 21, no. 7, pp. 17-26, 1986.

[Sarkar 89] V. Sarkar, *Partitioning and Scheduling Parallel Programs for Multiprocessors*, Research Monographs in Parallel and Distributed Computing. Cambridge, MA: The MIT Press, and London: Pitman Publishing, 1989.

[ShaWan 89] S.M. Shatz and J.-P. Wang (eds.), *Tutorial: Distributed Software Engineering*. Washington, D.C.: IEEE Computer Society Press, 1989.

[SkeGla 85] S. Skedzielewski and J. Glauert, "IF1–An Intermediate Form for Applicative Languages," Lawrence Livermore National Laboratory Report LLL/M–170, 31 July 1985.

[Starke 90] P.H. Starke, *Analyse von Petri-Netz-Modellen*, Leitfäden und Monographien der Informatik. Stuttgart: B.G. Teubner, 1990.

CHAPTER 6

Static Task Allocation and Code Generation

6.1 Dynamic versus Static Scheduling and Task Allocation

When the application program has been partitioned into tasks representing units of operations to be executed as a whole, these tasks have to be assigned to the processors for execution. Generally, this process is called scheduling, i.e., deciding when and where a task is to be executed with regard to the constraints imposed. Such constraints, are the demand on resources made by the tasks and and the availability of these resources from the processors. Some performance measure must be optimized, mostly either the system utilization (i.e., all components of the system should be busy) or the response time (i.e., a set of cooperating tasks should be processed in the shortest time). Examples of resources are memory space, peripherals, or means of communication.

In more specific terms, the assignment process can be split into two phases: the allocation phase and the scheduling phase (in a more restricted sense than mentioned above). Allocation deals with assigning the processor on which a task is executed, and scheduling handles the time of execution, i.e., the order of execution of the tasks allocated to one specific processor. This order is called a schedule of the tasks assigned to a processor.

It is obvious that scheduling and allocation are mutually dependent. Changes made in one phase affect the other. For example, if a schedule includes idle times because a task is delayed while another task blocks a specific resource, this access conflict can be resolved by moving the delayed task to another processor where the desired resource is free at that time. The allocation is thus modified to optimize the schedule.

In general-purpose computing systems, the needs of an application for computation power and resources are not known in advance. Scheduling and allocation have to be done at run-time, i.e., dynamically. Therefore, efficient and quick methods must be employed for this purpose, possibly sacrificing optimality in order not to consume too much computing power for "non-productive" work. These scheduling duties are normally performed by the operating system.

For real-time systems, this trade-off becomes one of the critical issues during the system design. Whereas in so-called hard real-time systems processing must have terminated by a given deadline and the tasks must have been computed correctly, the violation of either requirement is tolerated in soft real-time systems. In the latter, sometimes even a reduced quality of results can be accepted to meet computation deadlines ([LiLiSh 91]). A good survey of approaches taken for scheduling in hard real-time applications is given in [ChStRa 88]. The authors conclude that only dynamic scheduling is sufficiently flexible to satisfy all the demands in such a system.

When scheduling is done dynamically, both static and dynamic allocation of tasks is possible. One choice is static allocation which is a less expensive but less flexible solution. The other possibility is dynamic allocation which needs more computing power but is more generally applicable ([HaLee 91]). Dynamic scheduling algorithms can allow or disallow task pre-emption.

However, if the application satisfies certain conditions it is possible to determine statically an allocation and a schedule. Above all, it must be possible to identify at compile-time the resources needed (e.g., memory space) and the order and the number of invocations of the tasks. This condition is fulfilled by signal processing and control applications, as shown in Chapter 3.

If a set of tasks is executed repeatedly, either a non-periodic or a periodic schedule can be generated. In non-periodic schedules, no overlapping of the task set executions is allowed. Therefore, the schedule is determined on the assumption that the tasks are executed only once, hence the name "non-periodic schedule". Periodic schedules try to use overlapping in order to increase the processor utilization (processor load) to reduce the schedule length (i.e., to increase the rate of results generation). This topic is found in detail in [Agne 89], [KoKaTa 90], [ParMes 89], [ParMes 91], [CurMad 91].

The difference between a periodic and a non-periodic schedule for the same set of tasks is illustrated in the Gantt chart in Figure 6.1, with the associated processor loads shown in Table 6.1. The loads are computed according to equations 6.3 and 6.4 given in Section 6.2.2.

Table 6.1 *Processor Loads for Non-periodic and Periodic Schedules*

	Non-periodic Schedule Load	Periodic Schedule Load
Mean	28/42 = 0.667	28/33 = 0.848
Processor 1	8/14 = 0.571	8/11 = 0.727
Processor 2	10/14 = 0.714	10/11 = 0.909
Processor 3	10/14 = 0.714	10/11 = 0.909

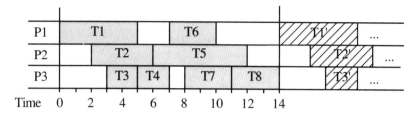

a) Non-periodic Schedule, Length 14

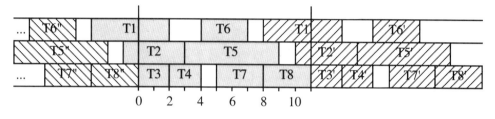

b) Periodic Schedule, Length 11

Figure 6.1 Non-periodic and Periodic Task Schedules

Using static allocation and scheduling has two advantages. Allocation and scheduling may be performed on a host computer which allows the use of better development tools. The application code is then transferred to the target system. By this means, considerable run-time overhead is saved, especially for real-time systems. Since limiting the expenses for allocation and for scheduling is not an issue with this approach, more expensive algorithms may be used. Better allocations and schedules are thereby generated.

If the task allocation is done in an intelligent way, scheduling becomes much easier. Then, many scheduling conflicts are avoided if not only the kind and the number of resources needed by the tasks is taken into account but also the time when they are claimed. One task utilizing one specific resource will then become ready only after the same resource has been released by another task placed on the same processor.

In order to save the expense of static scheduling, where the exact execution order of the tasks has to be fixed, a simple dynamic round-robin scheduler can be used, which activates the tasks as soon as they are ready. The exact times when the tasks execute are irrelevant for the scheduler. In addition, finding the precise start and termination times of the tasks would require sophisticated simulations of the program execution. As a consequence, it is no longer possible to give the exact execution time of the whole application (sometimes also called response time or task turnaround time). Only an estimate based on the utilization of the PEs is possible.

84 *Static task allocation and code generation*

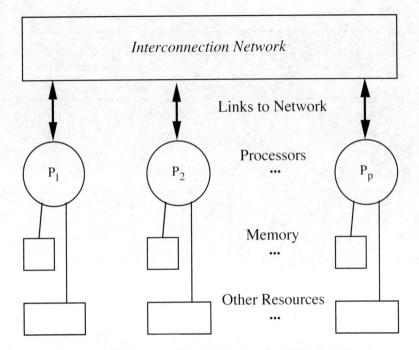

Figure 6.2 General Hardware Configuration for a Multicomputer

In the multicomputer used in this project, a simple round-robin scheduler executing very quickly has been microcoded in each processor by the manufacturer (see Appendix A). Therefore, only the static task allocation is performed.

6.2 The Static Task Allocation Problem

The static task allocation problem is generally formulated in terms of an optimization problem where a performance measure has to be optimized under a set of constraints. The general problem is outlined below, together with a model of the multiprocessor system assumed in most work on task allocation. For general distributed computing systems, a good introduction to the subject is given in [ChHoLa 80]. The problem outlined here applies to real-time data processing (see also [Borrma 86]).

6.2.1 General Machine Configuration

The computing systems treated consist of a set $P = \{P_1, P_2, ... P_p\}$ of p processors which can work independently. Each processor has its own private memory. The set $R = \{R_1, R_2, ..., R_r\}$ includes all resources, such as peripheral devices and communication equipment. In homogeneous multicomputers all processors are identical, whereas heterogeneous multicomputers consist of various types of processors. The processors are interconnected by a communication network which allows message passing between any two elements. Figure 6.2 shows the structure of the general multicomputer.

There are various ways of creating the interconnection network. In systems with few processors a bus system may be the cheapest solution. However, bus contention very soon becomes a problem ([Thoeni 88]). In larger systems, a fast serial network, e.g., Ethernet, is feasible. However, such a network also becomes congested under heavy load. Systems including many processors, such as Hypercube architectures, in most cases possess only few serial links per processor. Each processor is connected to neighbours. If the communication is directed to processors not directly connected, the messages are passed through intermediate processors. The communication delay then becomes a function of the distance of the communicating processes and may increase considerably. A communication cost model for such systems is given in [KräMüh 87].

6.2.2 General Static Task Allocation Problem

In general the problem is the distribution of a set of m tasks to p processors so that the system throughput is maximized (see [ChHoLa 80], [Stone 77]). Stone uses the term "module" for task or process. Throughput maximization is obtained by minimization of communication overhead. Therefore, if the amount of intermodule communication (IMC) is given for all pairs of processes, the interprocessor communication cost (IPC) is determined by the IMC and the allocation. The aim is to find an allocation of the tasks to the processors so that the sum of all interprocessor communication costs and task execution costs on the processors is minimal.

A precise analytical definition of the problem is given by Borrmann ([Borrma 86]). He also considers the aspect of graceful degradation by assigning a priority to each process. In a processor failure a new precomputed allocation is activated and as many processes as possible are reassigned. Each task j is assigned the priority w_j. The task execution costs do not appear explicitly in the equations. Instead the tasks are considered to use the resource "computing power". Therefore the execution costs are part of the equation for the resource constraints.

The following notation is used:

$P = \{1..p\}$ set of processors

$M = \{1..m\}$ set of tasks

$R = \{1..r\}$ set of resources

$W \subseteq P$ set of non-faulted processors

$A = [a_{ih}]$ task assignment matrix $a_{ih} \in \{0, 1\}$ $i \in M, h \in P$

               if task i is assigned to processor h then $a_{ih} = 1$

$B = [b_{ir}]$	resource demand matrix		$b_{ir} \geq 0 \; i \in M, r \in R$

$B = [b_{ir}]$ resource demand matrix $b_{ir} \geq 0 \; i \in M, r \in R$
task i needs b_{ir} units of the resource r (e.g., MFlops of computational power, or MBytes of memory)

$C = [c_{hr}]$ resource capacity matrix $c_{hr} \geq 0, \quad h \in P, r \in R$
processor h possesses c_{hr} units of the resource r

$V = [v_{ij}]$ communication volume matr. $|v_{ij}| \geq 0$ $i, j \in M \times M$
tasks i and j exchange v_{ij} bytes of data. v_{ij} is negative if task i and j must not be allocated on the same processor

$D = [d_{gh}]$ distance matrix $d_{gh} \geq 0$ $g, h \in P \times P$
the shortest path from processor g to processor h passes through d_{gh} intermediate processors

w_i priority of task i $w_i > 0 \; i \in P$
a low value of w_i indicates high priority

The Task Assignment Problem is defined as:

> Find an assignment A_{opt} of the tasks M to the processors P, considering the following conditions:

1. Objective Function 1 (OF 1) (equation 6.2):
Assignment of as many processes as possible according to their priority
Given the quality q of an allocation A

$$q(A) = \sum_{i=1}^{m} \sum_{j=1}^{p} \frac{1}{w_i} a_{ij} \qquad (6.1)$$

maximize this quality to find the optimum assignment A_{opt}:

$$q(A_{opt}) = \text{Max} \{ q(A) \} ! \qquad (6.2)$$

2. Objective Function 2 (OF 2) (equation 6.4):
Reduction of Interprocessor Communication (IPC)
The measure z of the communication costs of allocation A is defined in equation 6.2. Two tasks i and j exchange the amount v_{ij} of data when allocated on the processors k and l with distance d_{kl}.

$$z(A) = \sum_{j=1}^{m} \sum_{i=1}^{j} \sum_{k=1}^{p} \sum_{l=1}^{p} v_{ij} \, d_{kl} \, a_{ik} \, a_{jl} \qquad (6.3)$$

The static task allocation problem 87

Minimize the interprocessor communication z to find the optimum assignment A_{opt}, given that the quality $q(A)$ has its maximum value:

$$z(A_{opt}) = \text{Min} \{ z(A) \mid q(A) = \text{Max} \} ! \qquad (6.4)$$

3. Constraint 1 (C 1) (equation 6.9): *Load balancing*

The load u_j of processor j is defined as the ratio of the sum of all resources requested by the tasks on processor j to the value of the sum of all resources available on processor j:

$$u_j = \frac{\sum_{i=1}^{p} \sum_{l=1}^{r} b_{il}\, a_{ij}}{\sum_{l=1}^{r} c_{jl}} \qquad \text{load of processor j} \qquad (6.5)$$

The mean processor load $\bar{u}$ is defined as the ratio of all resources requested by all tasks to the value of the sum of all resources provided by all processors:

$$\bar{u} = \frac{\sum_{i=1}^{m} \sum_{l=1}^{r} b_{il}}{\sum_{\substack{j=1 \\ j \in W}}^{p} \sum_{l=1}^{r} c_{jl}} \qquad \text{mean processor load} \qquad (6.6)$$

A bound Δu can be introduced which depends on the mean load $\bar{u}$ as follows:

$$\Delta u = f(\bar{u}) = \begin{cases} 0 & \text{if } \bar{u} = 1 \\ \Delta u_{max} & \text{if } \bar{u} \ll 1 \end{cases} \qquad (6.7)$$

If the mean load $\bar{u}$ is one, i.e., the processors are fully utilized, the bound Δu is zero. If the mean load $\bar{u}$ is smaller, then an arbitrary bound Δu_{max} is taken. The load of processor j is not allowed to be larger than $\bar{u}+\Delta u$:

$$u_j - \bar{u} \leq \Delta u \qquad \forall\, j \in W \qquad (6.8)$$

4. Constraint 2 (C 2) (equation 6.8): *Resource utilization*

88 Static task allocation and code generation

The amount of resource r requested by all tasks on each processor must not exceed the capacity of the specific resource:

$$\sum_{i=1}^{m} b_{ir} a_{ij} \leq c_{jr} \qquad \forall j \in W, \forall r \in R \qquad (6.9)$$

5. Constraint 3 (C 3) (equation 6.10): *Avoidance of task coresidence*
No pair i, j of tasks may be placed on the same processor l if the intertask communication volume v_{ij} has a negative value:

$$\text{if } v_{ij} < 0 \text{ then } a_{il} a_{jl} = 0 \qquad \forall i, j \in M \times M, \forall l \in P \qquad (6.10)$$

6. Constraint 4 (C 4) (equation 6.11): *Avoidance of faulty processors*
No task may be placed on a defective processor:

$$\sum_{i=1}^{m} a_{il} = 0 \qquad \forall l \in P\text{-}W \qquad (6.11)$$

7. Constraint 5 (C 5) (equation 6.12): *Unique assignment*
Each task i may be placed on maximally one processor:

$$\sum_{l=1}^{n} a_{il} \in \{0, 1\} \qquad \forall i \in M \qquad (6.12)$$

6.2.3 Static Task Allocation Problem for Homogeneous Real-time Multicomputers

The issues of fault tolerance and graceful degradation are not considered in this book. All tasks are thus allocated. The quality q of the allocation then yields the constant value

$$q = \sum_{i=1}^{m} \frac{1}{w_i}$$

according to 6.1. It is independent of the allocation A. The objective function OF 1 is therefore satisfied in any case, as well as the constraint C 4.

In some approaches, the total cost of the application is the subject of the minimization. The total cost is composed of the intertask communication costs and the execution costs of the tasks on the processors. The tasks are thus forced to processors where their execution is cheapest. However, in a homogeneous system these costs are the same on all processors so there is no point in including these constant costs in the objective function to be minimized.

In the multicomputer assumed here, no communication network linking all processors directly to each other exists. Each processor possesses a limited number of serial links (four for the T800 Transputer) which connect two processors. The links are regarded as communication resources. The topology of this limited interconnection network is not previously defined. It results from the allocation phase. Furthermore, in order to limit the interprocessor communication costs, messages are forwarded only to processors connected directly, i.e., no message routing through intermediate processors takes place. Since the elements of the communication distance matrix D then can only take on the values 1 or 0, the matrix D degenerates to a connectivity matrix, indicating the connections among the processors.

The difficulty is that the cost of transferring messages becomes dependent on the allocation. While in a universal network the transmission costs ideally are proportional to the communication volume v_{ij}, communication between two tasks may become impossible because the tasks have been allocated to two processors not directly linked. Intertask communication is cheaper if the two tasks reside on the same processor than when they are located on different ones, but it is not totally free, as assumed in many papers in the literature. No direct connection exists when the available communication resources have already been assigned to other interprocess communication channels. However, it is possible to share links between two pairs of communicating tasks residing on the same pair of processors.

Handling resources becomes easier in a homogeneous multicomputer because there are no restrictions as to where to allocate certain tasks requiring special resources. Moreover, load balancing is eased by the homogeneity of the system.

Due to the advances of VLSI technology and the fact that signal processing applications require a storage area only of a known size, the memory size constraint has lost importance and is not considered here. No peripheral devices such as printers etc. are taken into account. The only remaining resource constraint is the interprocessor link configuration.

The aspect of load balancing has to be treated with caution in hard real-time systems. If an application program is repeated periodically, periodic schedules can be generated which maximize throughput and system utilization, thereby balancing the load of the processors. This involves some kind of pipelined execution of consecutive invocations of a task set, i.e., the first tasks of the next execution of the program are started before the last tasks have finished running, as mentioned in Section 6.1. Through this overlapping scheme much time that would otherwise be wasted can be saved.

One of the characteristics of control applications, however, is that the acquisition of new input values and the computation of the next output values can only start when the previous computation cycle has been terminated. This excludes any pipelined approach for most algorithms, and only non-periodic schedules are applicable. Inevitably, then, at the beginning and end of program execution only a few processors can be active, thus creating an unbalanced load. For this reason, minimizing the program execution time rather than load balancing is the primary goal.

90 Static task allocation and code generation

Exempt from this general rule are algorithms which are explicitly designed to have several stages, such as the Kalman filter, where the new state is estimated using old information and is updated as soon as the new measurements are available. A part of the computations can then be performed before the sampling instance, thus allowing some kind of pipelining. However, these algorithms are not treated in a special way since minimization of execution time is the primary goal in any case.

For a description of the modified task assignment problem some additional notation is required:

$U = \{1..u\}$ set of processors utilized in the final allocation, $U \subseteq P$

$O = [o_{ij}]$ communication overhead $o_{ij} \geq 0$ $i, j \in M \times M$
communication from task i to task j causes total costs of o_{ij}. These costs depend on the allocation of the tasks

$L = [l_{ij}]$ link matrix $l_{ij} \in \{0, 1\}$ $i, j \in M \times M$
if $l_{ij} = 1$ then task i sends data to task j

$TP = [tp_{gh}]$ task parallelism matrix $tp_{gh} \in \{\text{false, true}\}$
$g, h \in P \times P$
if tp_{gh} then the tasks g and h can be executed in parallel

The modified Task Assignment Problem for a homogeneous multicomputer then reads:

> Find an assignment A_{opt} of the tasks M to the processors P and the topology of the interprocessor communication network, considering the following conditions:

1. Objective Function A (OF A) (equation 6.15)

Reduction of the Interprocessor Communication (IPC)

The measure z of the communication costs of allocation A is defined below. Two tasks i and j result in o_{ij} cost for communication. The cost depends on the intertask communication volume v_{ij} and on whether the two tasks are allocated on the same processor or on different processors, as described in equation 6.14.

$$z(A) = \sum_{j=1}^{m} \sum_{i=1}^{j} \sum_{k=1}^{p} \sum_{l=1}^{p} o_{ij}\, a_{ik}\, a_{jl} \qquad (6.13)$$

with

$$o_{ij} = \begin{cases} v_{ij}\, T_{ext} + T_s & \text{if } a_{ik}\, a_{jk} = 0 \quad \text{(ext. comm.)} \\ v_{ij}\, T_{int} + T_s & \text{if } a_{ik}\, a_{jk} = 1 \quad \text{(int. comm.)} \end{cases} \qquad (6.14)$$

T_s : communication set-up time
T_{ext} : communication time per byte for external communication
T_{int} : communication time per byte for internal communication

Minimize the interprocessor communication I to find the optimum assignment A_{Opt}:

$$z(A_{Opt}) = \text{Min} \{ z(A) \} \; ! \tag{6.15}$$

2. Objective Function B (OF B) (expression 6.16):
Minimum program execution time

$$\text{Finishing time of last task of data flow graph} = \text{Min} \; ! \tag{6.16}$$

3. Constraint A (C A) (equation 6.17):
Maximum communication resource utilization

The maximum number of communication links available on any processor is l_{max}. For any processor j, the sum of bidirectional links (pairs of one entering and one leaving link) must not exceed l_{max}.

$$\left(\sum_{i=1}^{p} \frac{l_{ij} + l_{ji}}{1 + l_{ij}l_{ji}} \right) - l_{max} = 0 \qquad \forall \, j \in P \tag{6.17}$$

4. Constraint B (C B) (equation 6.18):
Avoidance of task coresidence of tasks executable in parallel

$$\text{if } tp_{ij} \text{ then } a_{il} \, a_{jl} = 0 \qquad \forall \, i, j \in M \times M, \, \forall \, l \in P \tag{6.17}$$

with

tp_{ij} = true if tasks i and j are executable in parallel

5. Constraint C (C C) (equation 6.20):
Maximum number of processors used

The number |U| of elements of the set U is the number of processors used, and the number |P| of elements of the set P is the number of processors available. With the set of unused processors

$$U \subseteq P \tag{6.19}$$

the number of unused processors has to be minimized:

$$|P| - |U| \geq 0 = \text{Min !} \tag{6.20}$$

The main differences in the formulation of the assignment problem for inhomogeneous multicomputers (as given by Borrmann, e.g.) are:

- Load balancing among the processors is not primarily sought
- The program execution time is required to be minimal
- Resource utilization maximization deals only with communication devices

The objective functions and constraints are discussed briefly in the following.

Because a full allocation of all tasks is sought, objective function A (OF A) contains no process priorities. The interprocessor communication is computed differently. Not only is communication between tasks on different processors considered, but also communication between tasks located on the same processor is taken into account. The exact communication cost model has been developed in Chapter 4. The distances between processors no longer appear since messages are forwarded only to those processors directly connected.

The second objective function introduced is the minimization of program execution time. This is the central issue of parallel processing for real-time systems. However, since only the task allocation is handled statically and scheduling is performed dynamically, the program execution time is difficult to measure in advance.

For this reason, the demand for minimal execution time cannot be a practical measure for evaluating the quality of an allocation. Instead, the two constraints A and B ((C A) and (C B)) are introduced. The link matrix L used in (C A) signifies: if any task located on processor i sends data to any task hosted by processor j, then l_{ij} equals one. Otherwise l_{ij} equals zero. The sum in (C A) indicates the number of links of a processor used unidirectionally or bidirectionally. This number must not exceed the number l_{max} of links available in hardware. By requiring the difference to be zero, all links have to be utilized. Thus maximal parallelism is sought.

Constraint B forbids the placement of two tasks on the same processor if they are potentially executable in parallel. This minimizes resource access conflicts and task switching by the scheduler. The scheduler has to activate only one task at a time, since by this allocation strategy the tasks placed on the same processor become ready at different times. Again, parallelism is preserved.

Constraint C requires that all processors be used for the execution of the program. Frequently this proves to be impossible due to the limited number of communication links. Therefore, the maximum value is sought which must not exceed the number of PEs available. In some cases it would not even make sense to enforce the use of all PEs since it is possible that the application program does not exhibit enough parallelism to keep all processors busy.

Constraints A and B are also likely to collide. When maximum parallelism is preserved by (C B), the number of links used quickly exceeds that of available links.

However, since the number of links available is limited by the hardware, respecting (C A) is more important than fulfilling (C B).

The three constraints (C A), (C B), and (C C) are necessary but not sufficient conditions for achieving minimum execution time as required by objective function (OF B). However, these constraints provide helpful guidelines for the pursuit of a good allocation of the tasks.

6.3 Solutions Proposed in the Literature

A very large number of solutions to the problem of task allocation (TA) has been proposed in the literature. The terminology used is inconsistent in that the problem is called "task assignment problem", "module allocation problem", "static scheduling problem", and variations thereof. The exact problem formulation differs, but most authors concentrate on minimizing either the sum of interprocessor communication and task execution costs or the program execution time.

However, it is possible to group the investigations according to the approaches taken.

Table 6.2 *Classification of Allocation Approaches Described in the Literature*

Graph Theoretic
  [Stone 77], [ChHoLa 80], [Bokhar 81], [SheTsa 85], [Bokhar 87],
  [Lo 88], [LeLeKi 92]
Numerical Optimization, Branch-and-Bound
  [RaChGo 72], [ChHoLa 80], [MaLeTs 81], [MaLeTs 82], [Sincla 87],
  [Fernán 89], [KoKaTa 90], [SheGag 91], [BiCoSu 92], [ShWaGo 92]
List Scheduling
  [AdChDi 74], [KasNar 84], [SarHen 86], [Towsle 86], [LoGli 87],
  [Thaler 87], [HwChAn 89], [Sarkar 89], [Löffle 90], [ThaMos 90],
  [CurMad 91]
Simulated Annealing
  [Steele 85], [BolMid 91], [BüEsMa 91], [DHoDev 91]
Clustering Methods
  [GylEdw 76], [ChHoLa 80], [Efe 82], [Campbe 85], [Borrma 86],
  [HoPaFe 86], [ChuLan 87], [LeeAgg 87], [LoGli 87], [MüGoKr 87],
  [KimBro 88], [BaxPat 89], [GeVeYa 90], [Housti 90], [GerYan 91]

One approach described in the literature is to find the optimum solution to the minimization problem. For solving this problem, graph theoretic approaches have been found, but they apply only to restricted cases.

Due to the NP-completeness of finding the optimum solution for the general case ([Bokhar 81], [Coffma 76]), most solutions described apply an heuristic method. They range from numerical optimization (implicit enumeration), list scheduling approaches, and simulated annealing to clustering methods where the interconnection structure of the task graph is used to form clusters of tasks which are allocated to the processors.

Table 6.2 lists the literature according to the approaches. The properties of the single approaches are pointed out in the following sections citing selected papers. A good survey of the older work is given by Borrmann ([Borrma 86]).

6.3.1 Graph Theoretic Approach

A method for finding the optimum solution for a restricted version of the Task Allocation Problem for systems with two processors has been developed by Stone ([Stone 77]). The task graph is augmented with two nodes, one per processor. From each processor node an edge is drawn to each task node. The edges are labelled with the sum of the task's communication and execution costs. The costs are computed assuming that the task is placed on the processor to which it is linked by that edge. With the maximum-flow/minimum-cut algorithm by Ford and Fulkerson ([ForFul 56]), the new graph is cut into two parts so that the edges crossing the boundary represent the minimum total cost. Lee, Lee, and Kim ([LeLeKi 92]) extend Stone's work for linear arrays of processors.

Since focusing only on the communication and execution costs inplies placing all tasks on one processor, Lo ([Lo 88]) introduced the concept of "interference costs". If two tasks need the same resources when placed on one processor, the execution cost is augmented by an amount representing this conflict. By driving apart tasks competing for the same resources, parallelism is preserved. Lo also extended the limitation of the problem to two processors by substituting one processor by a so-called supernode containing p-1 processors and iteratively applying the original algorithm to that configuration.

Although an attractive approach, this method becomes very expensive for a large number of tasks and processors and is difficult to extend to include additional constraints, e.g., on resource utilization.

A solution for task graphs which are tree-shaped and which can then be transformed into simpler structures is proposed by Bokhari ([Bokhar 81], [Bokhar 87]). Shen and Tsai ([SheTsa 85]) use a graph matching method called weak graph homomorphism to map the task graph onto the processors.

6.3.2 Numerical Optimization Approach

While it is impossible to examine all the potential assignments due to the rapid expansion of combinatorial work, strategies for partial enumeration originating from Operations Research are applicable.

Dynamic Programming and two related heuristic methods are utilized in [RaChGo 72] in order to precompute an optimum schedule, but without considering communication costs. However, due to the large complexity of the problem, only systems with two processors and 54 nodes at most are treated in the examples.

One approach described in [KoKaTa 90] uses a branch-and-bound technique with backward- and forward-searching algorithms from the theory of 0-1 integer programming. Sinclair ([Sincla 87]) employs the branch-and-bound method with underestimates. After a partial assignment has been achieved, the probable costs of all full assignments which include that partial assignment are estimated. According to these estimates it is decided whether to pursue further this branch of the search tree.

In [ShWaGo 92] the 0-1 integer programming problem is solved using the A* algorithm used in Artificial Intelligence. The authors put the emphasis in their approach more on maximizing the system's reliability than on minimizing execution costs.

Billionnet, Costa, and Sutter formulate the task allocation problem as the minimization of a quadratic pseudo-Boolean function with linear constraints. They use the Lagrangean dual problem to solve the linear program with a branch-and-bound method. They are able to handle large numbers of tasks by this approach, but they neglect intertask communication costs for tasks which reside on the same processor.

Fernández-Baca ([Fernán 89]) even proposes a polynomial-time algorithm for the assignment using nonserial dynamic programming, but only for task graphs which are partial k-trees.

The difficulty with these methods is to find an appropriate cost function which does not prematurely exclude potentially good solutions but which nevertheless limits the cases to be inspected to a tolerable number.

6.3.3 List Scheduling Approach

List scheduling methods are well known from dynamic schedulers used in operating systems. The name is based on the fact that a list of the tasks ready to execute is maintained (dynamically or statically). Each element of the list carries a tag indicating its priority, determined from deadline or other constraints. The list is sorted according to the priorities. The order of execution is thus fixed. The priority of a task may change over time in order to increase its chance of being executed (for the case of deadline-driven priorities). When performed dynamically, the trade-off is scheduling cost against accuracy. The better the priorities have to reflect the true relations among the tasks, the more their computation causes overhead.

Static list scheduling relies mostly on methods using the critical path (CP) of the task graph. The tasks of the critical path are allocated first so that no delays are introduced. All the other tasks are then fitted around this partial assignment according to the various constraints imposed. Since these tasks do not belong to the critical path, the total program execution time is not increased.

For the allocation of the tasks which do not belong to the critical path, several heuristic approaches have been proposed. Traditionally, they are given acronyms representing verbose descriptions. Adam, Chandy, and Dickins in their paper [AdChDi 74] compare the methods HLFET (Highest Level First with Estimated (execution) Times), HLFNET (Highest Level First with No Estimated Times), RANDOM (this speaks for itself), SCFET (Smallest Colevel First with Estimated Times), and SCFNET (Smallest Colevel First with No Estimated Times). They define the level of a task node as the longest path from that task node to an exit node of the graph, and the colevel as the longest path from an entry node of the graph to the same task node. Their favourite candidate is HLFET. This performs best because it possesses the most accurate information on the workload to be performed for completing the program's execution. In their experiments they found that HLFET remains within 5% of the optimal solution. However, Lo and Gligor ([LoGli 87]) compare HLFET to a so-called group scheduling approach (described in Section 6.3.5). They conclude that list scheduling is more sensitive to variations of computation parameters (degree and pattern of task interaction) when few processors are available and cannot completely match the performance of group scheduling for many processors.

CP/MISF (Critical Path/Most Immediate Successor First), proposed by Kasahara and Narita ([KasNar 84]), already tends towards the philosophy of the clustering methods described below by emphasizing the locality of communication. In contrast, ETF (Earliest Task First) ([HwChAn 89]) relies on the time a task is ready to produce a greedy scheduling.

The most critical issue in list scheduling approaches is computation of the critical path. In systems with a good communication network providing high bandwidth, the communication costs are not greatly influenced by the communication load, in contrast to architectures with few links (such as hypercubes). All the same, the communication costs are highly dependent on the mapping which is to be determined with the help of the critical path. But this critical path, in turn, is computed using these same communication costs. Therefore, good estimates of the critical path are difficult to obtain.

This problem can be eased by optimizing the schedule obtained by a CP method. Thaler ([Thaler 87], [ThaMos 90]) optimizes the completed schedule with respect to limited resources. Löffler ([Löffle 90]) iterates incremental scheduling and resource allocation steps in order to take into account the interdependence of scheduling and resource allocation.

For periodic schedules, Curtis and Madisetti ([CurMad 91]) use a wavefront analysis to identify the longest path, together with the respective resource and communication constraints.

6.3.4 Simulated Annealing Approach

Simulated annealing ([KiGeVe 82], [KiGeVe 83]) is becoming a popular method for avoiding getting trapped in local minima when solving complex optimization problems. Instead of following a steepest descent trajectory (as in conjugate gradient methods), the path is perturbed by random walks with decreasing probability.

Steele ([Steele 85]) applies simulated annealing to the allocation problem using an objective function that incorporates terms to model processor load and communication costs. Since communication costs are directly proportional to the distance between the tasks, they fail to account for congestion effects on single links.

Bütler, Esser, and Mattmann ([BüEsMa 91]) allocate the tasks representing a Petri Net on a given set of processors. The objective function is to minimize communication between the clusters. The authors do not assume any fixed interconnection topology, but configure the communication network after the allocation according to requirements.

Bollinger and Midkiff ([BolMid 91]) tackle the task allocation problem under the restriction that there is at most one task per processor. They use a two-step procedure. First they assign the tasks to the processors (task annealing), then they determine the communication paths among the processors (connection annealing) for a hypercube architecture. The objective function to be minimized contains the total communication cost and the maximum communication on a single link.

D'Hollander and Devis ([DHoDev 91]) form so-called annealing packets of ready tasks and assign one packet at the time. Their cost function is a sum of normalized communication and load-balancing terms. Their experiments showed equal or better performance than the HLFET list scheduling algorithm.

6.3.5 Clustering Approach

The principle of cluster analysis was originally applied to the analysis of statistical data gathered in biology, chemistry, and medicine (see, for example [Anderb 73], [Spaeth 85]). Task allocation, as in statistics, seeks to identify clusters of objects which have some kind of coherence and can therefore be treated as an entity. In most cases, these clusters are built from task nodes which have strong connections among each other, thus placing them on one processor eliminates much interprocessor communication cost.

A clustering is called nonlinear if two independent (i.e., not directly connected) tasks are mapped in the same cluster. Otherwise it is called linear. Figure 6.3 shows examples of nonlinear and linear clusterings of a task graph.

In the example for nonlinear clustering, task T2 contains two unconnected nodes. They could be processed in parallel if they were not placed in the same task. This shows that nonlinear clustering is unfavourable if maximum parallelism is sought. This fact has been confirmed by a comparison of clustering algorithms described in [GerYan 91].

Most authors use relatively crude models for the communication costs, where costs are at best proportional to the distance between processors and to the volume of communication. However, a few authors have formulated fairly sophisticated models. A detailed communication model for a hypercube structure set up by Lee and Aggarwal ([LeeAgg 87]) allows several communication channels to use the same hardware link at different times without increasing the overhead. For a homogeneous bus-based multiprocessor architecture using message passing, communication costs as a function of bus utilization are given by Houstis ([Housti 90]). However, all authors assume that intertask communication on the same processor is free.

98 *Static task allocation and code generation*

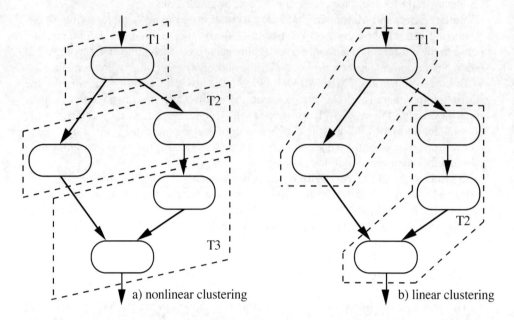

Figure 6.3 Examples of Nonlinear and Linear Clustering

Task assignment basing on cluster analysis is a two-step approach. First, an initial (possibly only partial) assignment of the task to the processors is made. In the subsequent steps, this initial solution is refined and optimized until no further improvement of the solution is found, or some constraints are violated, or the allowed expenses for the optimization are exceeded.

For the initial assignment, the simplest possibility is one task per processor ([GylEdw 76], [Housti 90]). A random assignment is not very effective since the first optimization steps would be wasted to correct gross misplacements. Therefore, most authors construct a fairly sophisticated initial assignment. This is mostly guided by the objective function to be minimized, sometimes disregarding some of the constraints ([ChHoLa 80], [Efe 82], [Borrma 86]). If communication costs are to be kept low (an aim that most authors pursue), then tasks which communicate heavily with each other are allocated on adjacent processors.

Instead of considering all task nodes at a time for allocation, Baxter and Patel ([BaxPat 89]) propose in their LAST algorithm (Localized Allocation of Static Tasks) following the graph's topology when selecting the next node to be allocated. Only nodes connected to other nodes already allocated are eligible for allocation. These nodes are subsequently placed on the processor of the predecessor to which they have the strongest connection (i.e., to the node where most of the incoming communication originates). A very similar idea is presented by Lee and Aggarwal ([LeeAgg 87]) and Hong, Payne, and Ferguson [HoPaFe 86].

For refining the initial assignment, all or only selected pairs of tasks are inspected for coresidence. The potential gain in the objective function is then calculated for the case where such a pair is separated again ([GylEdw 86], [KimBro 88], [Houstis 90]). Other schemes include exchange mechanisms for improving the first solution or to fulfil all constraints ([ChHoLa 80], [Efe 82], [ChuLan 87], [LeeAgg 87]). Borrmann ([Borrma 86]) even proceeds in three steps by first choosing single tasks, then pairs of tasks, and finally whole groups of tasks for exchange.

The fact that the task graph's topological properties (i.e., the connectivity) and the activation order of the tasks determine the sequence of allocation ensures that as few decisions as possible are made which have to be reversed. For this reason, the clustering approaches are mostly straightforward allocation methods with no costly backtracking options.

6.4 Two-phase Linear Clustering Approach

Minimizing interprocess communication is an important problem in message-passing multicomputers, and making effective use of the limited serial communication links is even more important for good performance. Therefore, the clustering heuristic was chosen to solve the task allocation problem. It seems the most appropriate of the methods described in the survey above.

With linear clustering, in each step of the optimization only potentially "good" candidates are considered for coresidence on the same processor. By placing two connected tasks on the same processor, at least one communication channel can be eliminated. This allows direct control of the constraint (C A) on the use of the communication resources mentioned in Section 6.2.3. An additional advantage of this local approach is that it is not necessary to evaluate the total communication cost formulated as objective function (OF A) at every step of the optimization. Finding the move which maximizes the decrease in the local communication cost (i.e., which minimizes a PE's communication cost) also guarantees the largest decrease of the total communication cost. The constraint (C B) which forbids coresidence of tasks executable in parallel is fully respected during the initial phase of linear clustering since only serially dependent tasks are placed on the same PE. However, when each PE hosts several tasks it may happen that two tasks are placed on the same processor which are both dependent on the same task but for themselves independent.

In an early phase of the work, experiments were conducted with nonlinear clustering. This showed no advantages over linear clustering, since only connected tasks could be merged on one PE. Moreover, the computational complexity was higher than with linear clustering and resulted in a significantly increased processing time.

Figure 6.4 shows the procedure for allocating the tasks on the processors. In Appendix B.5 the rules are specified in detail.

100 *Static task allocation and code generation*

<u>ALLOCATION</u> (fgr, alpha):

(* place each task of fgr on a PE of its own
compute the level of all tasks
(the longest path from entry point of the graph to the tasks)
determine the maximum number maxlinks of links used on any PE
determine the number excesspe of excess PEs in the system *)

excesslinks := maxlinks - numberoflinks_per_PE
while (excesspe>0) **or** (excesslinks>0) **do**
    if av_tasknumber<alpha **then**   (* PHASE 1 *)
        find the pair of PEs mergepair which fulfills the following
        conditions:
            - the communication between the PEs of mergepair
                costs much
            - joining the pair's processes on one PE eliminates the
                biggest number of links from the system
        <u>join processes</u>(mergepair)
    else (* av_tasknumber≥alpha *)   (* PHASE 2 *)
        for all processors PEi **do**
            <u>distribute processes</u>(PEi, processlist)
            (* place tasks to the PE they are most conn. to
                (lowest level first) *)
            determine the weighted sum of all links used in
                the system
            keep in minpe the PE which yields the lowest value of
                the sum
            <u>collect processes</u>(PEi, processlist)
        end for all
        <u>distribute processes</u>(minpe, processlist)
    end if
    determine the maximum number maxlinks of links used on any PE
    excesspe := excesspe - 1
    av_tasknumber := numberoftasks / maxpe
    excesslinks := maxlinks - numberoflinks_per_PE
end while
end allocation

Figure 6.4 Allocation of the Tasks on the Processing Elements

First, each task is placed on a (hypothetical) PE of its own. Normally, this number exceeds the number of PEs available in the system. Generally, more links than are available are needed between the PEs. Therefore, in each iteration one PE is eliminated from the system so that the number of links decreases monotonically. Since one PE is removed in each step, the algorithm will terminate in any case, at worst yielding the trivial solution with only one PE. The allocation itself proceeds in two phases as described below. Switching from phase one to phase two is controlled by the average number of tasks per PE, which initially has the value one. As soon as it reaches the threshold value entered by the user, phase two is started.

In an earlier version, the transition condition was the number of PEs. However, this had the disadvantage that for problems of different sizes the aim of processing switched at different levels of processing. For large problems, processing was almost ended before phase two was begun, while it had hardly started for small ones. When choosing the average number of tasks, for any problem size the number of PEs has decreased by the same ratio. Phase two is then begun at a state of processing where sufficient room is left to pursue the aims of phase two.

Finally, each PE has a list of the links associated with it. Through these lists the topology of the interconnection network among the PEs is defined.

6.4.1 Reducing the Number of Interconnections

So far, each edge in the data flow graph was created by a link of its own. Early experiments with allocation showed very quickly that this results in so many communication channels that there is hardly a chance of finding a useful allocation

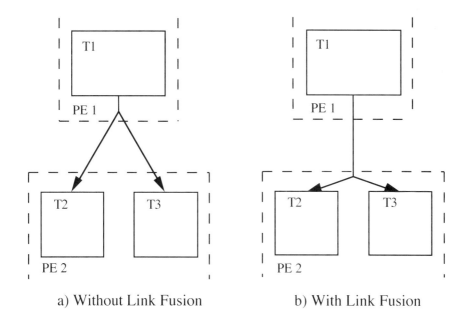

Figure 6.5 Interconnection Elimination By Link Fusion

102 *Static task allocation and code generation*

meeting the hardware constraints. Another observation was that with algorithms consisting of vector operations many elements of an array often have to be distributed from or gathered in one task. Many channels then leave from or enter the same task. This observation led to the development of the concept of link fusion illustrated in Figure 6.5.

Link fusion is applicable if data are sent from one source on PE 1 to multiple sinks located on PE 2, but not necessarily in the same task. Then, instead of sending the same data items over multiple links from PE 1 to PE 2, only one link is used for the transmission between the PEs involved, and data are distributed among the consumers on the PE.

Through this simple concept, a considerable number of links can be eliminated without introducing a large amount of overhead.

The other method for eliminating links is by time multiplexing. The outputs from nodes in any process on PE 1 are sent through a multiplexer task and transmitted over one link to PE 2 where the inputs waiting for the values are located. Link multiplexing is

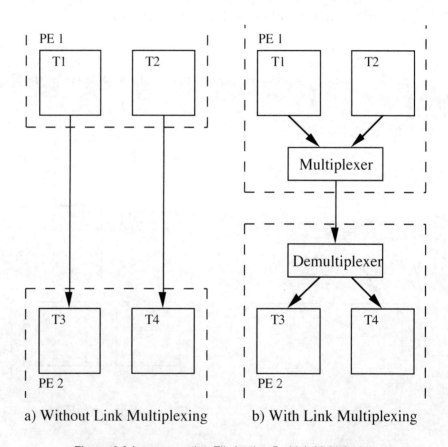

Figure 6.6 Interconnection Elimination By Link Multiplexing

shown in Figure 6.6.

The introduction of multiplexer and demultiplexer tasks causes some additional computation costs and introduces delays in the transmission. All the same, it is worth paying the price, since many links can be saved by this simple method. Additionally, the chance is small that two tasks on a PE have to communicate over a multiplexed link at the same time since they are linearly dependent on each other and execute one after the other. Messages are therefore not delayed by colliding access to links due to multiplexing.

These optimizations are applicable independently for both directions of the bidirectional physical links. Currently, the combination of the link fusion and link multiplexing optimizations for one monodirectional link is not implemented. Priority is given to link fusion since it causes less run-time costs.

6.4.2 Phase One: Clustering Heavily Communicating Tasks

Phase one aims at a quick reduction in the number of processors in the system and the removal of the communication links with the highest cost. For this purpose, in each iteration a list of all communicating pairs of PEs is built, sorted according to the interprocessor communication costs. Among those in the top half (i.e., the pairs with high communication costs) the pair is identified which offers the greatest saving of links when the task sets of the two PEs are merged onto one PE. One of the two processors is eliminated from the system and its tasks (initially there will be only one task on it) are moved to the other PE. The links eliminated are not only those connecting the two PEs directly. The possibility of eliminating links leading from other processors to the PE pair to be merged is also investigated. The techniques for further reducing the number of links are link fusion and link multiplexing, as outlined above.

While the objective is to lower the total number of links used in the system, no attention is paid to the situation on individual processors. That means that PEs with many excess links are not more likely to be considered for merging than those with spare links, unless their communication costs are high. All the same, by removing links with high communication costs, the aims of the objective function (OF A) from Section 6.2.3 are followed. By considering only half the number of PE pairs, the selection proceeds quickly.

6.4.3 Phase Two: Matching Interconnection Topology and Communication Resources

Phase two is activated as soon as the average number of tasks per PE has reached the threshold value which is typically set to two to four. In this final stage of the allocation meeting both hardware constraints (C A) and (C C) is the main aim. These constraints concern the communication resource utilization and the utilization of as many of the available processors as possible, but not more. Therefore a more extensive search is performed for finding a solution which satisfies the physical constraints of the system while sacrificing as little performance (parallelism and communication time) as possible.

Continuing to merge task sets of PEs would create a very unbalanced load. For this reason, the tasks of a PE are distributed to the other processors to which they are connected most in terms of communication costs. This helps to distribute the load as evenly as possible. The task clusters are modified but still remain linear because only connected tasks are merged.

In each iteration of the algorithm all PEs are examined. When the tasks of one PE are distributed, all links used in the system are counted as in phase one. However, this time the weighted sum of the difference between the number *lno* of links actually used and the number *noflinks* of links available is determined. The formula applied to each PE is

$$\text{new sum} = \begin{cases} \text{old sum} + 5 \times (lno - noflinks) & \text{if } lno > noflinks \\ \text{old sum} + (noflinks - lno) & \text{if } lno \leq noflinks \end{cases}$$

Excess links are thus counted fivefold whereas spare links count only once. The specific values for the weights were determined by experiments and emerged as the best compromise of removing excess links and using spare ones. Through these weights, the method which eliminates the largest number of excess links is chosen. The PE with the lowest value of the sum associated with it is selected for the final distribution of its tasks.

Due to the fact that in each iteration one PE is removed from the system and excess links are eliminated with priority, the search converges towards a solution which satisfies all the hardware constraints and still offers the best performance possible.

6.4.4 Computational Complexity Analysis of the Allocation

Determining the amount of work to be done during the allocation phase is as difficult as it was for partitioning. While the parameter alpha indicating the average task number is easy to include in the calculation, much depends on the configuration of the actual solution found. For a worst-case estimation it has to be assumed that the trivial solution with only one PE results. All other solutions require fewer computations.

The figure to start with is the number m of PEs after partitioning the data flow graph. This number m is in the range of N/2 ... N/3, where N denotes the number of nodes in the expanded graph. Additional variables used are the number n_i = m ...(1) of PEs in the system and the average number of processes pr_i = 1 ... alpha ... (m) per PE in iteration i.

The operation counts for the single actions during phases one and two are listed in Table 6.3. The figures given are the worst-case estimation of the operation count for the allocation process. It is assumed that processing has to continue until only one PE is left in the system. In any case, phase two is dominant as far as the workload is concerned.

The terms of fourth and third order grow so fast that if the number m of PEs after partitioning exceeds just a few, all other terms become negligible.

It is surprising that the threshold value alpha of average number of tasks does not appear in the final sum. However, this is due to the fact that in phase one a simple clustering method is used which does not contribute significantly to the total complexity.

As seen in the partitioning step, the number m of PEs is mostly in the range of N/2 ... N/3 for an expanded graph with N nodes. This leads to the complexity for the allocation phase of $O\left(\frac{N^4}{81} + \frac{N^3}{27}\right)$ which approaches $O(N^3)$ if N is less than 100.

Table 6.3 Computational Complexity of the Allocation

Function Performed:	Operation Count	Total
Phase One:	$\lceil m(1-1/alpha)\rceil$ Iterations	
build list of PE pairs	n_i	
determine number of links before merging	n_i	
determine number of links after merging	n_i	
delete list of PE pairs	n_i	
Total per Iteration:	$4 n_i$	
Total Phase One:		$2 m^2 (1 - 1/alpha^2)$
Phase Two:	$\lceil (m - alpha)\rceil$ Iterations	
distribute the processes	$n_i(pr_i^2 + pr_i^3)$	
determine sum of excess/spare links	$n_i^2 \, pr_i$	
collect processes	$n_i \, pr_i^3$	
Total per Iteration:	$pr_i(n_i + n_i^2) + pr_i^2 \, n_i + pr_i^3 \, n_i$	
Total Phase Two:		$m^4 + 4 m^3 - alpha^4/2 - alpha^3$
Grand Total (approximately):		$m^4 + 4 m^3$

6.5 Translating the Partitioned and Allocated Data Flow Graph into Target Code

In the preceding steps the data flow graph has been partitioned into tasks. These tasks have been allocated to the PEs so that the constraints imposed by the hardware are fulfilled. After these processing stages, the data flow graph contains all the information necessary for translating it into source code for the target machine. Each node is associated with a process which in turn belongs to a specific processing element. Each edge is classified as external channel among two PEs, as internal channel connecting two processes on the same PE, or as variable linking two nodes of the same process.

The exact characteristics of the target system have so far been irrelevant. The only data about the target multiprocessor used are the communication cost model, the node execution cost table, and the information on the maximum number of processors available. If another type of multiprocessor is used, only the cost tables have to be replaced. The principle of processing is not changed by such modifications.

For the translation of the data flow graph into source code, the exact syntax of the target system's programming language is of no relevance. The only condition is that constructs for sending and receiving messages must be available, together with the possibility of placing processes for concurrent execution on the same PE.

Therefore, the language constructs for all these functions and those for realizing the functions of the nodes are kept separate from the actual translation program. It is thus theoretically possible to switch the target language without changing the translator itself.

6.6 Generating OCCAM Code

Several projects ([GauLee 88], [SchWüs 88], [Shield 88]) have been described in which SISAL was ported to the Transputer, i.e., where the data flow graphs described in the Intermediate Form 1 (IF1) were translated to OCCAM.

However, since full use of the capabilities of SISAL was allowed, great difficulties arose in their implementation on the Transputer. One obstacle is SISAL's dynamic use of memory which is not supported by OCCAM. While this can be partly overcome by a memory-managing process ([SchWüs 88]), SISAL's functional recursion conflicts with the Transputer's static process model. While Gaudiot and Lee ([GauLee 88]) do not comment on the subject, Schibli and Wüst ([SchWüs 88]) and Shield ([Shield 88]) come to the (correct) conclusion that these features are unimplementable on the Transputer, as presumed in the project described here. All three projects provide some kind of dynamic process management.

As described in detail in Chapter 2, for real-time signal processing, dynamic process creation and dynamic data structures are neither necessary nor desirable due to the high run-time overhead they introduce. For this reason, only static allocation of the computations is implemented which is easily realizable in OCCAM.

The nodes of the data flow graph are contained in a second data structure apart from the connections through the edges. In this linked list, each PE carries a list of the processes allocated on it and of all its links. In turn, each process possesses a list of the graph nodes it comprises. Therefore, it is possible to convert the data flow graph to OCCAM code in a single pass.

Two files are generated. The first, *occam.all*, contains the PE declarations. There, as required in OCCAM, the type of the PEs is declared, together with the placement of the channels on the physical links and with the headers of the processes allocated to the PEs.

The source code generated from the data flow graph is written to the second file, *occam.src*. For each PE, a main process is created which handles data communication over the physical links. If fused or multiplexed links are used, the appropriate protocols and multiplexer/demultiplexer processes are created. In this main process, all the tasks residing on the PE are placed in a *PAR* statement, i.e., the microprogrammed scheduler runs them in parallel as soon as they are ready to execute.

The translation process is described in detail in Appendix B.6. Here, only the principle of the language-independent translation of the graph nodes is presented.

The file *target_language.txt* contains the corresponding OCCAM code sequence for each IF1 graph node and for all other constructs which are needed (communication, process declaration). Since the function of some nodes is dependent on the type of its inputs, several code sequences may exist for one node. In the sequences, dummy variable names $i, with i any integer, are present which have to be replaced before the code can be inserted into the final program. As an example, the code for the Plus node (node identifier 141) is shown in Figure 6.7.

The different cases are labelled @*case_number* and reach to the next line beginning with an "@" character. For the Plus node, case one is given if the inputs are Boolean values; then an OR operation is performed. The ordinary addition is realized for integer values by the PLUS operator, whereas for real or double real operands the "+" sign is inserted.

For the compound nodes which represent while-loops and if-clauses, several code segments are provided for the initialization, body, and returns sections.

Through this concept of separating the translator program from the actual syntax of the target language, modifications of the code generator are easily possible.

```
_141      %Plus
@001       %boolean
$1 := ($2 OR $3)
@002       %integer
$1 := $2 PLUS $3
@003       %real, double
$1 := $2 + $3
@
```

Figure 6.7 OCCAM Code for the Plus Node

6.8 References

[AdChDi 74]   T.L. Adam, K.M. Chandy and J.R. Dickson, "A Comparison of List Schedules for Parallel Processing Systems," *Comm. of the ACM*, vol. 17, no. 12, pp. 685-690, 1974.

[Agne 89]   R. Agne, "Zur Scheduling-Problematik in Echtzeitsystemen," Universität Kaiserslautern, Bericht 2/89 des Zentrums für Rechnergestützte Ingenieursysteme, 1989.

[Anderb 73]   M.R. Anderberg, *Cluster Analysis for Applications*, Probability and Mathematical Statistics Series, vol. 19. New York, London: Academic Press, 1973.

[BaxPat 89]   J. Baxter and J.H. Patel, "The LAST Algorithm: A Heuristic-Based Static Task Allocation Algorithm," in: *Proc. Int. Conf. Parallel Proc.*, Aug. 1989, pp. II-217–II-222.

[BiCoSu 92]   A. Billionnet, M.C. Costa and A. Sutter, "An Efficient Algorithm for a Task Allocation Problem," *J. of the ACM*, vol. 39, no. 3, pp. 502-518, 1992.

[Bokhar 81]   S.H. Bokhari, "A Shortest Tree Algorithm for Optimal Assignments Across Space and Time in a Distributed Processor System," *IEEE Trans. Softw. Engineering*, vol. 7, no. 6, pp. 583-589, 1981.

[Bokhar 87]   S.H. Bokhari, *Assignment Problems in Parallel and Distributed Computing*. Boston, Dordrecht, Lancaster: Kluwer Academic Publishers, 1987.

[BolMid 91]   S.W. Bollinger and S.F. Midkiff, "Heuristic Technique for Processor and Link Assignment in Multicomputers," *IEEE Trans. Computers*, vol. 40, no. 3, pp. 325-333, 1991.

[Borrma 86]   L. Borrmann, *Allokation von Rechenprozessen in verteilten Realzeitsystemen*, VDI Fortschrittberichte, Reihe 10: Informatik/Kommunikationstechnik, Nr. 59. Düsseldorf: VDI Verlag, 1986.

[BüEsMa 91]   B. Bütler, R. Esser, and R. Mattmann, "A Distributed Simulator for High Order Petri Nets", in: G. Rozenberg (ed.), *Advances in Petri Nets 1990*, LNCS vol. 483, Berlin, Heidelberg, a.o.: Springer-Verlag, 1991, pp. 47-63.

[Campbe 86]   M.L. Campbell, "Static Allocation for a Data Flow Multiprocessor," in: *Proc. Int. Conf. Parallel Proc.*, Aug. 1985, pp. 511-517.

[ChStRa 88]   S.-C. Cheng, J.A. Stankovic and K. Ramamritham, "Scheduling Algorithms for Hard-Real-Time Systems – A Brief Survey," in: *Hard Real-Time Systems (Tutorial)*, J.A. Stankovic and K. Ramamritham (eds.). Washington D.C.: IEEE Computer Society Press, 1988.

[ChHoLa 80]   W.W. Chu, L.J. Holloway, M.-T. Lan, *et al.*, "Task Allocation in Distributed Data Processing," *Computer*, vol. 13, no. 11, pp. 57-69, 1980.

[ChuLan 87]   W.W. Chu and L.M.T. Lan, "Task Allocation and Precedence Relations for Distributed Real-Time Systems," *IEEE Trans. Computers*, vol. 36, no. 6, pp. 667-679, 1987.

[Coffma 76]   E.G. Coffman, *Computers and Job-Shop Scheduling Theory*, New York: John Wiley & Sons, Inc., 1976.

[CurMad 91]   B.A. Curtis and V.K. Madisetti, "Task Scheduling Super-compilers for the Georgia Tech Digital Signal Multiprocessor," submitted to *IEEE Trans. SP*, 18 September, 1991.

[DHoDev 91]   E.H. D'Hollander and Y. Devis, "Directed Taskgraph Scheduling Using Simulated Annealing," in: *Proc. Int. Conf. Parallel Proc.*, Aug. 1991, pp. II-180–II-185.

[Efe 82]   K. Efe, "Heuristic Models of Task Assignment Scheduling in Distributed Systems," *Computer*, vol. 15, no. 6, pp. 50-56, 1982.

[Fernand 89]   D. Fernández-Baca, "Allocating Modules to Processors in a Distributed System," *IEEE Trans. Softw. Engineering*, vol. 15, no. 11, pp. 1427-1436, 1989.

[ForFul 56]   L.R. Ford and D.R. Fulkerson, "Maximum Flow Through a Network," *Can. J. Math.*, vol. 8, pp. 399-404, 1956.

[GauLee 88] J.-L. Gaudiot and L.-T. Lee, "Occamflow: A Methodology for Programming Multiprocessor Systems," *J. Parallel and Distr. Comput.*, vol. 7, no. 1, pp. 96-124, 1988.

[GeVeYa 90] A. Gerasoulis, S. Venugopal and T. Yang, "Clustering Task Graphs for Message Passing Architectures," in: *Proc. 1990 Int. Conf. Supercomputing*, 11-15 June 1990. Amsterdam, New York: ACM Press, 1990, pp. 447-456.

[GerYan 91] A. Gerasoulis and T. Yang, "A Comparison of Clustering Heuristics for Scheduling DAGS on Multiprocessors," Report LCSR-TR-169, 20 September 1991, Department of Computer Science, Rutgers University, New Brunswick, NJ.

[GylEdw 76] V.B. Gylys and J.A. Edwards, "Optimal Partitioning of Workload for Distributed Systems," in: *Proc. IEEE COMPCON*, Fall 1976, pp. 353-357.

[HaLee 91] S. Ha and E.A. Lee, "Compile-Time Scheduling and Assignment of Data-Flow Program Graphs with Data-Dependent Iteration," *IEEE Trans. Computers*, vol. 40, no. 11, pp. 1225-1238, 1991.

[HoPaFe 86] Y.-C. Hong, T.H. Payne and L.B.O. Ferguson, "Graph Allocation in Static Dataflow Systems," *Computer Architecture News*, vol. 14, no. 2, pp. 55-64, 1986.

[Houstis 90] C.E. Houstis, "Module Allocation of Real-Time Applications to Distributed Systems," *IEEE Trans. Softw. Engineering*, vol. 16, no. 7, pp. 699-709, 1990.

[HwChAn 89] J.-J. Hwang, Y.-C. Chow, F.W. Anger, *et al.*, "Scheduling Precedence Graphs in Systems with Interprocessor Communication Times," *SIAM J. Comput.*, vol. 18, no. 2, pp. 244-257, 1989.

[KasNar 84] H. Kasahara and S. Narita, "Practical Multiprocessor Scheduling Algorithms for Efficient Parallel Processing," *IEEE Trans. Computers*, vol. 33, no. 11, pp. 1023-1029, 1984.

[KimBro 88] S.J. Kim and J.C. Browne, "A General Approach to Mapping of Parallel Computations upon Multiprocessor Architectures," in: *Proc. Int. Conf. Parallel Proc.*, Aug. 1988, pp. III-1–III-8.

[KiGeVe 82] S. Kirkpatrick, C.D. Gelatt and M.P. Vecchi, "Optimization by Simulated Annealing," IBM Research Report RC 9355, 1982.

[KiGeVe 83] S. Kirkpatrick, C.D. Gelatt and M.P. Vecchi, "Optimization by Simulated Annealing," *Science*, vol. 220, no. 4598, pp. 671-680, 1983.

[KoKaTa 90] K. Konstantinides, R.T. Kaneshiro and J.R. Tani, "Task Allocation and Scheduling Models for Multiprocessor Digital Signal Processing," *IEEE Trans. ASSP*, vol. 38, no. 12, pp. 2151-2161, 1990.

[KräMüh 87] O. Krämer and H. Mühlenbein, "Mapping Strategies in Message Based Multiprocessor Systems," in *Proc. PARLE 87, Parallel Architectures and Languages Europe*, Eindhoven, 1987, Vol. 2: Parallel Languages, Lecture Notes in Computer Science, vol. 259, Berlin, a.o.: Springer-Verlag, 1987, pp. 213-225.

[Lee 91] E.A. Lee, "Static Scheduling of Data-Flow Programs for DSP," in: *Advanced Topics in Data-Flow Computing*, J.-L. Gaudiot, L. Bic (eds.). Englewood Cliff, NJ: Prentice Hall, Inc., 1991, pp. 501-526.

[LeeAgg 87] S.Y. Lee and J.K. Aggarwal, "A Mapping Strategy for Parallel Processing," *IEEE Trans. Computers*, vol. 36, no. 4, pp. 433-441, 1987.

[LeeMes 87] E.A. Lee and D.G. Messerschmitt, "Static Scheduling of Synchronous Data Flow Programs for Digital Signal Processing," *IEEE Trans. Computers*, vol. 36, no. 1, pp. 24-35, 1987.

[LeLeKi 92] C.-H. Lee, D. Lee and M. Kim, "Optimal Task Assignment in Linear Array Networks," *IEEE Trans. Computers*, vol. 41, no. 7, pp. 877-880, 1992.

[LiLiSh 91] J.W.S. Liu, K.-J. Lin, W.-K. Shih, *et al.*, "Algorithms for Scheduling Imprecise Computations," *Computer*, vol. 24, no. 5, pp. 58-68, 1991.

[Lo 88] V.M. Lo, "Heuristic Algorithms for Task Assignment in Distributed Systems," *IEEE Trans. Computers*, vol. 37, no. 11, pp. 1384-1397, 1988.

[Löffle 90] C. Löffler, *Contributions to Architectural Design in Digital Signal Processing*, Series in Microelectronics, vol. 6, Konstanz: Hartung-Gorre Verlag, 1990.

[LoGli 87] S.P. Lo and V.D. Gligor, "Properties of Multiprocessor Scheduling Algorithms," in: *Proc. Int. Conf. Parallel Proc.*, August 1987, pp. 867-870.

[MaLeTs 81] P.Y. Ma, E.Y.S Lee and M. Tsuchiya, "On the Design of a Task Allocation Scheme for Time-Critical Applications," in: *Proc. IEEE Real-Time Systems Symposium*, 1981, pp. 121-125.

[MaLeTs 82] P.Y. Ma, E.Y.S. Lee and M. Tsuchiya, "A Task Allocation Model for Distributed Computing Systems," *IEEE Trans. Computers*, vol. 31, no. 1, pp. 41-47, 1982.

[MüGoKr 87] H. Mühlenbein, M. Gorges-Schleuter and O. Krämer, "New Solutions to the Mapping Problem of Parallel Systems: The Evolution Approach," *Parallel Computing*, vol. 4, no. 3, pp. 269-279, 1987.

[ParMes 89] K.K. Parhi and D.G. Messerschmitt, "Fully-Static Rate-Optimal Scheduling of Iterative Data-Flow Programs via Optimum Unfolding," in: *Proc. Int. Conf. Parallel Proc.*, August 1989, pp. I-209–I-216.

[ParMes 91] K.K. Parhi and D.G. Messerschmitt, "Static Rate-Optimal Scheduling of Iterative Data-Flow Programs via Optimum Unfolding," *IEEE Trans. Computers*, vol. 40, no. 2, pp. 178-195, 1991.

[RaChGo 72] C.V. Ramamoorthy, K.M. Chandy and M.J. Gonzalez, "Optimal Scheduling Strategies in a Multiprocessor System," *IEEE Trans. Computers*, vol. 21, no. 2, pp. 137-146, 1972.

[SarHen 86] V. Sarkar and J. Hennessy, "Compile-Time Partitioning and Scheduling of Parallel Programs," *SIGPLAN Notices*, vol. 21, no. 7, pp. 17-26, 1986.

[Sarkar 89] V. Sarkar, *Partitioning and Scheduling Parallel Programs for Multiprocessors*, Research Monographs in Parallel and Distributed Computing. London; Cambridge, MA: The MIT Press, and London: Pitman Publishing, 1989.

[SchWüs 88] P. Schibli and U. Wüst, "Erzeugung von OCCAM aus IF1-Datenflussgraphen," EE Diploma Thesis, Computer Engineering and Networks Laboratory, Swiss Federal Institute of Technology, Zurich, January 1988.

[ShWaGo 92] S. M. Shatz, J.-P. Wang and M. Goto, "Task Allocation for Maximizing Reliability of Distributed Computer Systems," *IEEE Trans. Computers*, vol. 41, no. 9, pp. 1156-1168, 1992.

[SheGag 91] T. Shepard and J.A.M. Gagné, "A Pre-Run-Time Scheduling Algorithm for Hard Real-Time Systems," *IEEE Trans. Softw. Engineering*, vol. 17, no. 7, pp. 669-677, 1991.

[SheTsa 85] C.-C. Shen and W.-H. Tsai, "A Graph Matching Approach to Optimal Task Assignment in Distributed Computing Systems Using a Minimax Criterion," *IEEE Trans. Computers*, vol. 34, no. 3, pp. 197-203, 1985.

[Shield 88] D.T. Shield, "Translating SISAL into OCCAM," M.Sc. Thesis, Dept. of Computer Science, University of Manchester, 1988.

[Sincla 87] J.B. Sinclair, "Efficient Computation of Optimal Assignments for Distributed Tasks," *J. Parallel and Distr. Comput.*, vol. 4, pp. 342-362, 1987.

[Spaeth 85] H. Spaeth, *Cluster Dissection and Analysis: Theory – FORTRAN Programs – Examples*, Computers and their Applications Series. Chichester: Ellis Horwood; Chichester a.o.: Halsted Press, 1985.

[Steele 85] C.S. Steele, "Placement of Communicating Processes on Multiprocessor Networks," California Institute of Technology, Computer Science Dept., Technical Report 5184:TR:85, April 1985.

[Stone 77] H.S. Stone, "Multiprocessor Scheduling with the Aid of Network Flow Algorithms," *IEEE Trans. Softw. Engineering*, vol. 3, no. 1, pp. 85-93, 1977.

[Thaler 87]   M. Thaler, *Analyse und Synthese von parallelen Signalprozessor-Architekturen*, Dissertation ETH Nr. 8240. Zürich, 1987.
[ThaMos 90]   M. Thaler and G.S. Moschytz, "A Data Flow Technique for the Efficient Design of a Class of Parallel Non-Data Flow Signal Processors," *IEEE Trans. ASSP*, vol. 38, no. 12, pp. 2162-2173, 1990.
[Thoeni 88]   U.A. Thoeni, "Enhancing the Processing Power of Real-Time VMEbus Systems," in: *Proc. Int. Conf. VMEbus in Research*, Zurich, 11-13 October 1988, C. Eck and C. Parkman (eds.), Amsterdam a.o.: North-Holland, 1988, pp. 477-486.
[Towsle 86]   D. Towsley, "Allocating Programs Containing Branches and Loops Within a Multiple Processor System," *IEEE Trans. Softw. Engineering*, vol. 12, no. 10, pp. 1018-1024, 1986.

CHAPTER 7

Case Studies

For most of the types of algorithms described in Chapter 3, examples were formulated in SISAL and run through the parallelization system. In this chapter, starting with a small example of a scalar product the single processing steps are shown in Section 7.1. In Section 7.2, digital filters (FIR and IIR type) of different sizes are investigated. Section 7.3 shows examples of linear state space controllers with observer, whereas in Section 7.4 the parallelization of a nonlinear controller is described. The distribution of a Runge-Kutta numerical integration algorithm on several processors is shown in Section 7.5. The chapter concludes with examples of fast Fourier transforms and a discussion of the results.

For all experiments, the configuration of the system was kept constant, i.e., eight PEs at most could be used, with four bidirectional links each.

Rating an allocation is a difficult task since the optimum solution is unknown. Therefore, for each example a series of runs with different values of the allocation parameter was performed. For each run, the resulting estimated execution time was compared to the algorithm's serial execution time to determine the PSR (parallel-to-serial execution time ratio.)

Since it is almost impossible to determine the exact parallel execution time if the PEs host more than one process each, the maximum load of any PE was assumed to represent the estimated total execution time for a particular distribution of an algorithm.

Depending on the load distribution, this time may be shorter than the real execution time, but it is a minimum bound of the execution time. If this time is longer than the required iteration time of the real-time application, another parallelization of the algorithms must be sought by choosing another value of the allocation parameter.

7.1 A Small Example

In order to illustrate the single steps performed during the parallelization, the computation of the scalar product of two vectors z and u with two elements each is shown as an example, where z is computed from another vector, x.

114 Case studies

The equation for the scalar product was given in equation 3.63. In order to show the effect of the parameter steering the function substitution, vector z is computed from another vector, x. The elements of z are computed by the formula $z_i = a \times x_i + b$. The other vector is $u = [5.0, 6.0]^T$. The call of the transform named *lin* is substituted by the function itself if the parameter in question is greater than two, since the function's graph contains two nodes.

In the header of the function *controlalgorithm*, the additional parameter *NumberOfInputs* appears which indicates the number of elements in vector x. Its value is given in function *main* where *controlalgorithm* is called. This function *main* is only used for the parallelization; it does not appear in the final OCCAM code as described in

```
function lin(y: real returns real)

let
  a := 2.0;
  b := 4.0
in
  a*y + b
end let
end function

function controlalgorithm(x: array[real];
        NumberOfInputs: integer returns real)

let
  u := array [1: 5.0, 6.0]
in
  for j in 1, 2
    R := u[j] * lin(x[j])
  returns
    value of tree sum R
  end for
end let

end function

function main(input: array[real] returns real)

  controlalgorithm(input, 2)

end function
```

Figure 7.1 SISAL Code of the Scalar Product

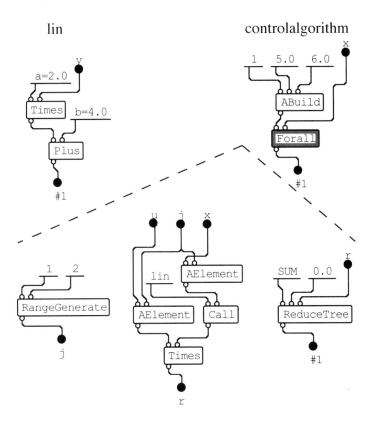

Figure 7.2 Data-Flow Graph of the Scalar Product

Chapter 4. Figure 7.1 shows the SISAL code describing the functions *lin* and *controlalgorithm*.

In Figure 7.2 the data-flow graph generated from this program is presented. It contains one Forall node in which the two multiplications of the scalar product are performed independently. In the first subgraph (shown left) the set {1, 2} of indices is generated. For each element of this set one element is selected from the vectors x and u, and the two values are multiplied as shown in the subgraph placed in the middle. Subsequently, the two partial results are summed using a tree reduction scheme (subgraph on the right).

In the next step, the communication volume of each edge in the graph is analysed. Since it is possible to determine the size of the input vector x with the auxiliary parameter in the call of *controlalgorithm*, the dimensions of all other arrays are determinable as well. In this example, however, all other data structures are of the simple data type Real.

The graph is then expanded. Figure 7.3 shows the expanded graph where the calls to

116 *Case studies*

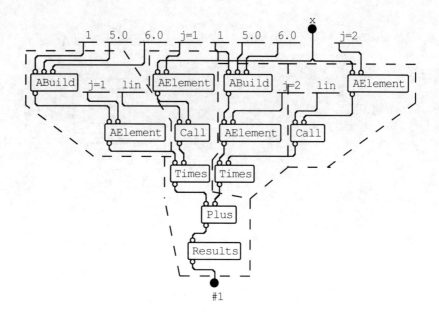

Figure 7.3 Expanded Graph Without Function Replacement

function *lin* are not replaced by the function's graph, while in Figure 7.4 all function calls have been eliminated. Replacing calls by the function code saves overhead at run-time, but can increase the number of nodes in a graph considerably, depending on the program structure. Thus the computation time for the parallelization may rise significantly.

Subsequently, the execution costs are determined. For all edges, external communication is assumed, i.e., data are transferred over interprocessor links. The cost of each node is determined from the processor-specific cost table according to the data type of the operands. The expanded graph is then partitioned into tasks. The tasks are marked in Figures 7.3 and 7.4 by dashed lines.

Table 7.1 shows the results of the partitioning and allocation steps. The acronym PSR again stands for parallel-to-serial execution time ratio. Initially, the serial execution time is determined by summing the computation time of all nodes. Each cycle lasts 50 ns for a T800 Transputer with a 20 MHz clock rate. For this example the serial execution time amounts to 587 instruction cycles if the function calls are not replaced, but only 293 cycles with the function body inserted. This large difference is due to the function call overhead and to the fact that the function itself contains only little work.

The simulation of the execution with an unbounded number of PEs (10 and 12 PEs actually used) yields execution times of around 700 cycles for both cases. In this phase each node of the data flow graph is regarded as a task of its own. Since each PE hosts only one process, the scheduler does not execute the processes in a time-sliced mode. Thus the simulation yields the exact execution time.

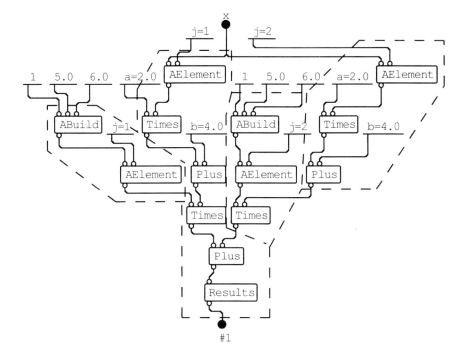

Figure 7.4 Expanded Graph With Function Replacement

Table 7.1 *Execution Times of the Partitioned Scalar Product*

function expansion	0 (no)	10 (yes)
serial exec. time	587	293
parallel exec. time	751	679
# nodes	10	12
partitioned exec. time	650	503
# tasks (# PEs used)	4	4
mean PE utilization	64.0%	68.0%
PSR	1.11	1.72
mean PE load	175.75	102.25
min. PE load	80	80
max. PE load	281	134
allocated, # PEs used	4	4

After partitioning the data flow graph into tasks, in both cases four tasks are formed. The simulated execution time is then 650 and 503 instruction cycles, respectively. For this small example, these times are longer than those resulting from computing the algorithm in a serial manner.

```
VAL lar IS 1(INT) :
CHAN OF REAL32 lnk.0 :
CHAN OF REAL32 lnk.2 :
CHAN OF REAL32 lnk.3 :
CHAN OF [2]REAL32 lnk.1 :

PLACED PAR

  PROCESSOR 1 T8

    PLACE lnk.0 AT Link0.out :

    pe1 (lnk.0)

  PROCESSOR 2 T8

    PLACE lnk.1 AT Link0.out :
    PLACE lnk.0 AT Link1.in :
    PLACE lnk.2 AT Link2.in :

    pe2 (x, out.0, lnk.1, lnk.0, lnk.2)

  PROCESSOR 3 T8

    PLACE lnk.3 AT Link0.in :
    PLACE lnk.2 AT Link1.out :

    pe3 (lnk.3, lnk.2)

  PROCESSOR 4 T8

    PLACE lnk.1 AT Link0.in :
    PLACE lnk.3 AT Link1.out :

    pe4 (lnk.1, lnk.3)
```

Figure 7.5 Allocation File of the Parallel Scalar Product

It is interesting to note that for the cases in which the function call is replaced by the function itself the mean PE load is lower and the load distribution more even among the PEs. This again is caused by the elimination of the costly function calls. The mean PE utilization (total execution time/PE load) of over 64% is very high for this kind of non-pipelined periodic execution of a program. It lies above the values to be expected in most cases.

In this special case, each PE is assigned only one process and therefore the simulated

A small example   119

execution time is exact. Generally, however, it is more difficult to determine the true termination time for systems involving multiple PEs executing several processes each. The shortest execution time possible is given by the maximum load of any PE in the system. Assuming that the load is distributed evenly among the PEs, the maximum load of any PE is a good indicator of the execution time found in reality.

Figure 7.5 shows the allocation file *occam.all* which is generated from the partitioned and allocated graph. It contains the link definitions and the distribution of the processes on the PEs. Each PE i contains one main process *pei* which in turn is composed of the processes representing the tasks of the partitioned data flow graph.

In Figure 7.6 the OCCAM source code for each PE is shown where all function calls have been replaced by the function's code. The process k of PE i is named *proci.k*. Since in this (atypical) case each PE hosts only one process, a SEQ statement (sequential execution) is placed at the end of each main PE process. Normally, all the processes on a PE are placed for parallel execution with a PAR statement.

```
PROC pe1(CHAN OF REAL32 lnk.0)

  PROC proc1.1(CHAN OF REAL32 lnk.0)

    REAL32 v.1 :

    SEQ
      VAL []REAL32 v.u.0 IS [5.0(REAL32), 6.0(REAL32)] :
      v.1 := v.u.0[1 MINUS lar]
      lnk.0 ! v.1
  : -- proc1.1

  SEQ
    proc1.1 (lnk.0)
: -- pe1

PROC pe2([2]REAL32 x, VAL REAL32 out.0,
         CHAN OF [2]REAL32 lnk.1,
         CHAN OF REAL32 lnk.0,
         CHAN OF REAL32 lnk.2)

  PROC proc2.1([2]REAL32 x, VAL REAL32 out.0,
               CHAN OF [2]REAL32 lnk.1,
               CHAN OF REAL32 lnk.2,
               CHAN OF REAL32 lnk.0)

    REAL32 v.0 :
    REAL32 v.1 :
    REAL32 v.2 :
    REAL32 v.y.3 :
```

```
    REAL32 v.4 :
    REAL32 v.5 :

    SEQ
      lnk.1 ! x
      v.y.3 := x[1 MINUS lar]
      v.2 := 2.0(REAL32) * v.y.3
      v.1 := v.2 + 4.0(REAL32)
      lnk.0 ? v.4
      v.0 := v.4 * v.1
      lnk.2 ? v.5
      out.0 := v.0 + v.5
    : -- proc2.1

    SEQ
      proc2.1 (x, out.0, lnk.1, lnk.2, lnk.0)
: -- pe2

PROC pe3(CHAN OF REAL32 lnk.3,
      CHAN OF REAL32 lnk.2)

  PROC proc3.1(CHAN OF REAL32 lnk.3,
        CHAN OF REAL32 lnk.2)

    REAL32 v.0 :
    REAL32 v.r.2 :
    REAL32 v.3 :

    SEQ
      VAL []REAL32 v.u.1 IS [6.0(REAL32), 5.0(REAL32)] :
      v.0 := v.u.1[2 MINUS lar]
      lnk.3 ? v.3
      v.r.2 := v.0 * v.3
      lnk.0 ! v.r.2
    : -- proc3.1

    SEQ
      proc3.1 (lnk.3, lnk.2)
: -- pe3

PROC pe4(CHAN OF [2]REAL32 lnk.1,
      CHAN OF REAL32 lnk.3)

  PROC proc4.1(CHAN OF [2]REAL32 lnk.1,
        CHAN OF REAL32 lnk.3)
```

```
  [2]REAL32 x :
  REAL32 v.0 :
  REAL32 v.y.1 :
  REAL32 v.2 :

  SEQ
    lnk.1 ? x
    v.y.1 := x[2 MINUS lar]
    v.0 := 2.0(REAL32) * v.y.1
    v.2 := v.0 + 4.0(REAL32)
    lnk.3 ! v.2
: -- proc4.1

  SEQ
    proc4.1 (lnk.1, lnk.3)
: -- pe4
```

Figure 7.6 Program Text of the Parallel Scalar Product

It is not useful to create this small algorithm with basically just four multiplications and three additions in a parallel implementation, as the execution times shown in Table 7.1 prove. However, the example was not intended to demonstrate the acceleration (in fact, none has been achieved), but to explain the single processing steps. Additionally, it illustrates the importance of good and prudent parallelization in order not to spend more time communicating data than can be saved by performing the computations in parallel.

7.2 Digital Filters

Discrete systems which receive input values at a fixed rate and compute output values using the input values and past internal state values are said to belong to the class of digital filters. Therefore, PID controllers (Section 3.3) fall within this category as well as lowpass filters, as described in Section 3.6.

However, two types of filters can be identified: the Finite Impulse Response (FIR) filter and the Infinite Impulse Response (IIR) filter. PID controllers belong to the group of IIR filters since they use the last output value for the computation of the following one, whereas in FIR filters only the past input values are utilized.

Since the computational structure of these two kinds of filters differs, examples of each are treated in separate sections.

7.2.1 FIR Filters

The general form of a Finite Impulse Response filter is given in equation 3.30:

122 Case studies

$$u_k = \sum_{i=0}^{q} b_{q-i}\, e_{k-i}$$

where the output u_k only depends on the past inputs e which are multiplied by the coefficients b.

For the order q of the filter five values were chosen: 10, 25, 50, 75, and 100. In the following, these examples are called FIR10, FIR25, and so forth.

Table 7.2 contains all the characteristic data found when partitioning the algorithm. The serial time is the pure computational load when executing the algorithm serially, without any communication costs. Below, the number of nodes in the expanded graph is given, followed by the simulated execution time of the partitioned graph. Here it is assumed that an unlimited number of PEs is available and that each PE executes only one task. The number of tasks formed is shown in the row beneath. Since no hardware constraints are respected in the partitioning phase, the last row shows the maximum number of links used on any PE.

Table 7.2 *Partitioning Data of the FIR Filters*

example	FIR10	FIR25	FIR50	FIR75	FIR100
serial time	2'247	3'469	7'473	11'697	16'048
# nodes	43	105	207	309	412
partitioned exec. time	6'405	33'585	130'764	283'529	503'802
# tasks	22	54	106	158	210
max. # links used	12	27	52	77	102

The figures in Table 7.2 show that the granularity, i.e., the size of the tasks, is very small because the number of tasks is quite close to half that of the nodes in the graph.

In Figure 7.7, the acronym PSR again stands for parallel-to-serial ratio, i.e., the ratio of the serial execution time and the estimated parallel execution time.

While the serial execution time rises proportionally to the problem size, the partitioned execution time increases almost proportional to the square of the problem size. This is due to the fact that for the partitioned execution time, all communication is accounted for. The communication itself is proportional to the number of edges in the graph, and this number increases according to the square of the number of nodes.

For the allocation of the small FIR filter with just 10 taps shown in Figure 7.7, the number of PEs decreases with increasing values of the allocation parameter. Therefore, the mean PE load increases. However, the maximum PE load varies greatly. The execution time found for values of the allocation parameter of 1.4...5.5 is low, with the minimum at 2.8...3.1. For these values, the estimated execution time amounts to only 0.33 of the serial execution time, and four PEs are used. The load is distributed almost evenly to the PEs with the minimum and the maximum PE load close to the mean load of all PEs.

For the FIR filter with 25 taps the results of the allocation runs are shown in Figure 7.8. Again, the numbers of PEs used decrease with increasing allocation parameters. The optimum solution is found in the range of the parameters of 9.5...10.5, with the minimum value of the PSR of 0.65, using four PEs.

The results for the 50-tap FIR filter shown in Figure 7.9 display a somewhat atypical behaviour. At first, for low parameter values, the number of PEs declines as the parameter value rises, reaching a minimum PSR of 0.62 at values of 10...18, using five

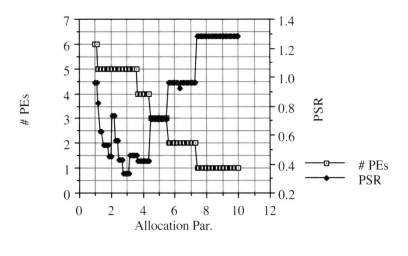

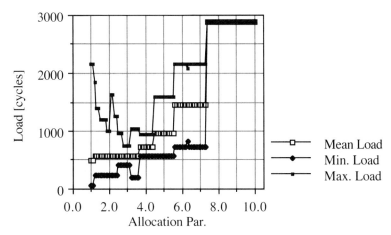

Figure 7.7 Allocation of the FIR10 Filters

124 *Case studies*

or three PEs. Then, for parameter values of over 30, the number of PEs rises to four again, yielding a PSR of 0.65. There, however, the difference in the load of the single PEs is large. The difference is smallest for parameter values of 20...24, but with a PSR of only 0.83. This illustrates once again the conflicting aims of good PE utilization and low execution time.

Figure 7.10 displays the results of the allocation of the FIR75 example of an FIR filter. The PSR is favourable for almost all parameter values, and reaches its minimum

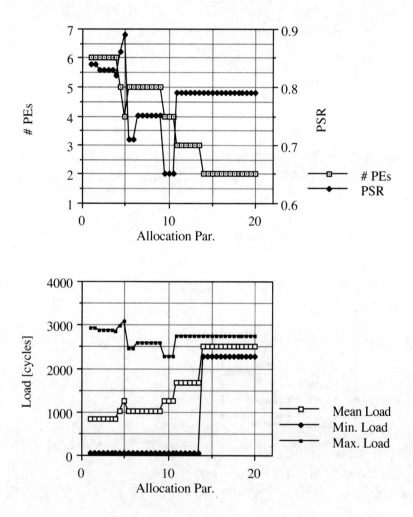

Figure 7.8 Allocation of the FIR25 Filters

of 0.61 for the parameters in the range of 20...26, where five PEs are utilized. However, the value of 0.64 for the parameters of 32...38 are almost as good, but there only three PEs are used.

For the large FIR filter FIR100, again only small values of the allocation parameter were tested (Figure 7.11). The results show that the parallelism is preserved relatively well with four to six PEs left active in the system, with the PSR close to 0.9.

The general observation emerging from these examples is that for the lowest values of the allocation parameter the number of PEs used is highest (seven or six) but always

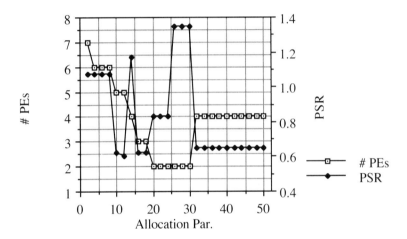

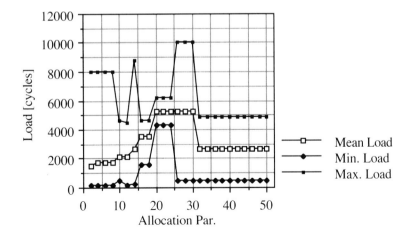

Figure 7.9 Allocation of the FIR50 Filters

126 *Case studies*

less than the number of PEs in the system (eight). Surprisingly, the PSR never falls below 0.33.

7.2.2 IIR Filters

Since Infinite Impulse Response filters use the feedback of past output values, their difference equation consists of two sums. The expression is (equation 3.33)

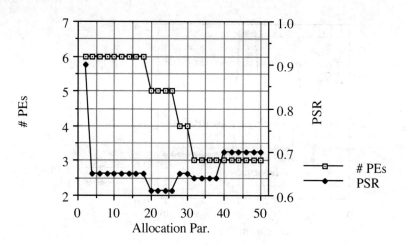

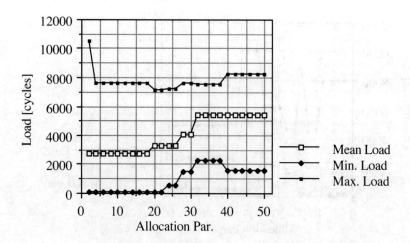

Figure 7.10 Allocation of the FIR75 Filters

$$u_k = \sum_{i=0}^{q} b_{q-i} e_{k-i} - \sum_{i=1}^{n} a_{n-i} u_{k-i} = b^T e_{0q} - a^T u_{1n}$$

Compared to the FIR filter, an additional coefficient vector a is used for the multiplication of the past outputs u.

The most general case is given when q = n. The following values were chosen for the respective examples IIR10, IIR25, and IIR50: 10, 25, 50.

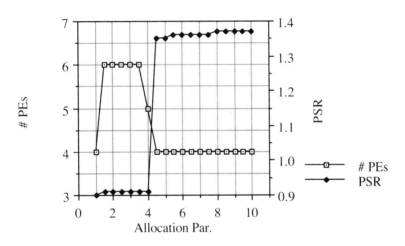

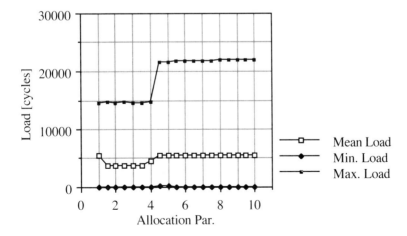

Figure 7.11 Allocation of the FIR100 Filters

128   Case studies

Table 7.3 displays data about the data-flow graphs of the example and the partitioning of the graphs.

The results of the allocation of an IIR filter with 10 state variables are displayed in Figure 7.12. The number of PEs used decreases steadily with increasing allocation parameters until only one PE is used. The difference between the maximum and minimum PE load gets smaller, but the PSR finally reaches 1.24. The best value of 0.6 results for a parameter value of 3. For parameter values from 20 to 40, the PSR remains at 1.24 with one PE used (not shown in the diagram).

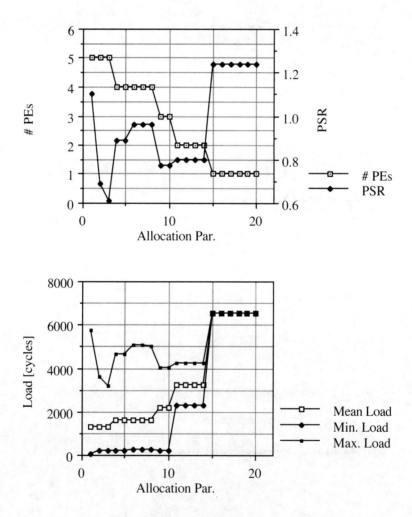

Figure 7.12 Allocation of the IIR10 Filters

Table 7.3 *Partitioning Data of the IIR Filters*

example	IIR10	IIR25	IIR50
serial time	5'240	8'674	18'332
# nodes	90	214	418
partitioned exec. time	13'602	69'597	261'135
# tasks	44	108	212
max. # links used	22	52	102

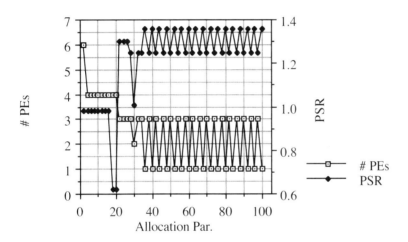

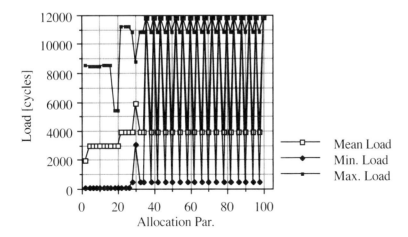

Figure 7.13 Allocation of the IIR25 Filters

132   *Case studies*

For the LQRE2 example shown in Figure 7.16, the best PSR achieved is 0.81 for three or two PEs. Load is distributed almost equally on the two PEs for the allocation parameter values 20...26. The reason is that the number of scalar products to be computed is a multiple of four. Therefore, these computations can easily be divided into two parts.

As for the LQRE1 example, the minimum for the ratio of the number of processes to the allocation parameter is found to be around 4.5. For parameter values over 26, the PSR rises to values above 1. Due to increased communication costs between the tasks

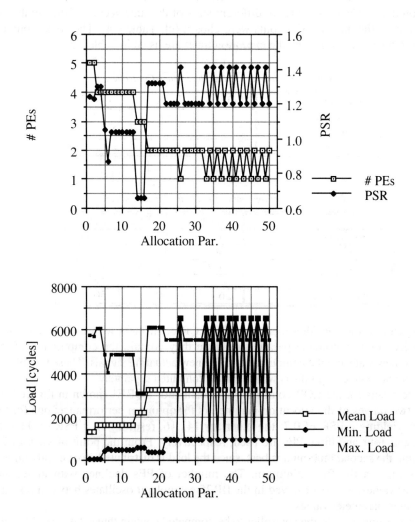

Figure 7.15 Allocation of the LQRE1 Controller

and the additional overhead for task administration, the parallel execution on one PE then lasts longer than the serial computation.

The largest example LQRE3 exhibits the same behaviour as the other LQR controllers. Its minimum PSR of 0.64 is found for a parameter value of 15, using four PEs. For high parameter values, the estimated execution time rises above the serial execution time.

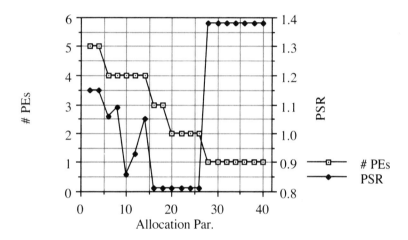

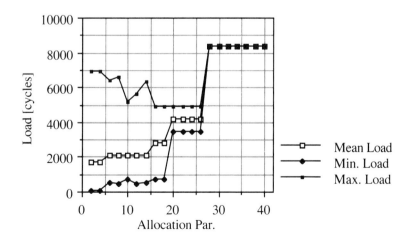

Figure 7.16 Allocation of the LQRE2 Controller

134 *Case studies*

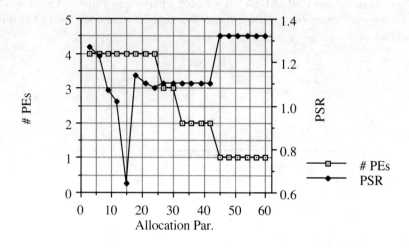

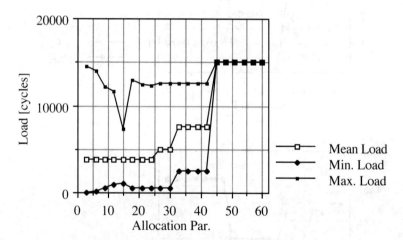

Figure 7.17 Allocation of the LQRE3 Controller

7.4 Nonlinear Controllers

As an example of a controller algorithm which utilizes nonlinear functions, the control law for stabilizing the speed of a DC motor was chosen. (The example is taken from M.F. Weilenmann and H.P. Geering, "Fast Nonlinear Control of Servo Drives with Current Saturations," *Proc. 13th World Congress on Computation and Applied*

Mathematics, 22-26 July 1991, Dublin, Ireland, pp. 1190-1191.) The basic controller is a PID controller with an additional anti-reset-windup extension and a limiter for the current. Its structure, together with the second-order plant, is shown in Figure 7.18.

The main controller (MC) is a conventional state feedback controller, and the auxiliary controllers (AC) consist of PI controllers. The auxiliary controllers limit the velocity x_2 of the plant. The desired values x_{2min} and x_{2max} are computed using the DC motor's position x_1 by the formulas

$$r \times \sqrt{2 \times b \times u_{min} \times x_1} \quad \text{and} \quad -r \times \sqrt{2 \times b \times u_{max} \times x_1}$$

respectively. The factors r and b are a shaping factor and a plant dependent value.

The special feature of algorithms of this kind is that they contain data dependent branches. The decision node and the partial graphs of all alternatives are always mapped to the same process. The reason is that only one of the parallel paths representing the alternatives is executed. Therefore, it does not make sense to place the alternatives on different processors.

Table 7.6 contains the data of the graph representing the algorithm.

Table 7.6 *Partitioning Data of a Nonlinear Controller*

serial time	1'227
# nodes	44
partitioned exec. time	2'946
# tasks	23
max. # links used	15

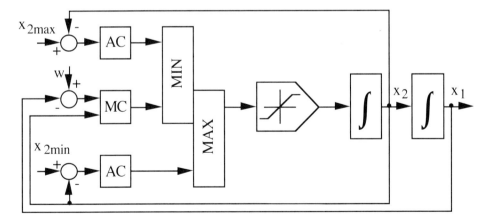

Figure 7.18 Structure of the Nonlinear Controller

136 *Case studies*

The results shown in Figure 7.19 for the nonlinear controller indicate that this application algorithm is unsuited for parallelization. Not only does it contain little work (less than 3000 instruction cycles), but this work is also contained to a great extent in the decision nodes which cannot be executed in parallel. For this reason, the load is distributed unevenly, and for most of the values of the allocation parameter the PSR lies above 1. Only for parameter values of 2 and 3 is the parallel execution time less than the serial execution time, with a minimum SPR of 0.64, and six PEs used. For higher parameter values, again an oscillation between one and two PEs can be observed.

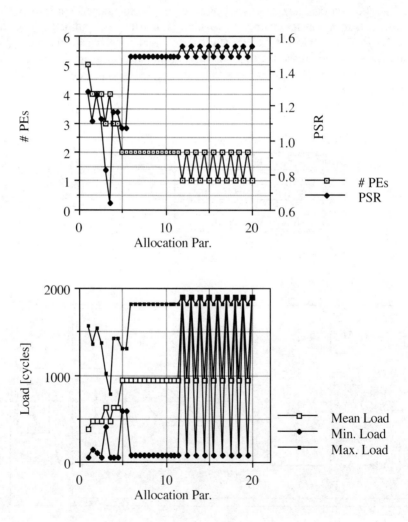

Figure 7.19 Allocation of the Nonlinear Controller

7.5 Numerical Integration

For numerical integration, the fourth-order Runge-Kutta method (equations 3.25a and b) was chosen, applied to the second-order differential equation of a pendulum $\ddot{\phi} + \nu^2 \sin\phi$. The sine function itself is evaluated monolithically on the processor by a combination of table lookup and interpolation, and is not subject to parallelization.

In the test runs RK4_1 and RK4_2, the nonlinear function of the system to be integrated was accessed by calls and by inserted code, respectively. The characteristic values of the graphs are shown in Table 7.7.

Again it becomes clear how much overhead is caused by the function calls compared to the serial execution time of the graph where the function code has been inserted.

Table 7.7 *Partitioning Data of the Runge-Kutta Integration*

example	RK4_1	RK4_2
serial time	7'204	5'048
# nodes	117	130
partitioned exec. time	6'599	5'511
# tasks	49	58
max. # links used	26	13

The RK4_1 example shown in Figure 7.20 displays behaviour similar to that of the IIR filters insofar as there are two ranges (4.4...7.8, 9.6...11.6) of the allocation parameters in which the PSR reaches a minimum of 0.42 and 0.57, respectively. The main difference, though, is that for the parameter values investigated, no significant increase in the PSR is observed for allocations with two PEs and rising allocation parameters. In this upper range, the computations are partitioned evenly among the two PEs. For parameter values from 15.8 to 20, the same solution with one PE resulted every time.

However, the best solution with four PEs is found for a parameter value of around 6, yielding a PSR of 0.42.

In the RK4_2 example depicted in Figure 7.21, parallelization is favoured by the fact that the function calls have been replaced by the function body. The number of tasks is increased, and the call and communication overheads are greatly reduced.

Here, the range of the allocation parameter for which the best parallel allocation is found is broader. In the best solutions found, three to six PEs are utilized, giving PSRs of 0.44, 0.5, and 0.58.

From the comparison of examples RK4_1 and RK4_2, the advantages of function call elimination become apparent: elimination of call and communication overhead, finer granularity (more processes of smaller size). An optimum solution is found for a broader range of the allocation parameter.

138 *Case studies*

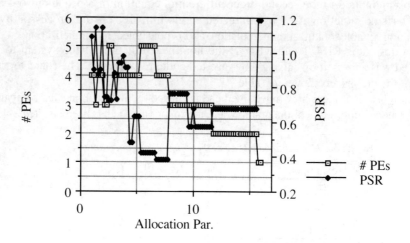

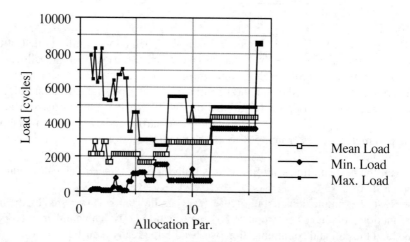

Figure 7.20 Allocation of the RK4_1 Numerical Integration

7.6 Fast Fourier Transform

For the fast Fourier transform as given in equation 3.45, two different transform sizes were chosen for the examples. FFT12 is a 12-point transform composed of radix-3 and radix-4 transforms, while FFT24 additionally contains radix-2 transforms.

The case FFT12_1 denotes no function expansion at all, while in FFT12_2 the small

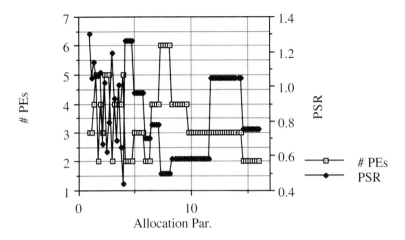

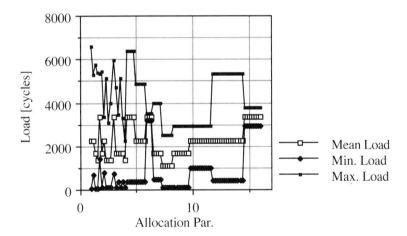

Figure 7.21 Allocation of the RK4_2 Numerical Integration

functions for the index computations were inserted. FFT12_3 is the full expansion of the two stages of the transform.

FFT24_1 denotes no function expansion of the 24-point transform. In FFT24_2, all three computational stages were expanded. Table 7.8 contains the data relating to the graphs and their partitioning.

These examples of the FFTs show the limitations of the current parallelization system. For those cases where no or only a small function expansion is performed, the partitions represent the calls of the single stages of the transform. However, each stage

140   *Case studies*

is performed as a whole on a single PE. Therefore, no speed-up is achieved and only three or four PEs are used.

As soon as the calls of the basic transforms are fully expanded, the number of nodes rises very rapidly. Since during partitioning only small processes with typically only two nodes are built, the number of processes to be allocated is still almost 1000 or even higher. As the number of operations of partitioning the graph and of allocating the nodes is proportional to the fourth power of the number of nodes, the computation time rises to levels which are beyond realizable values (estimated values: order of tens of days on a

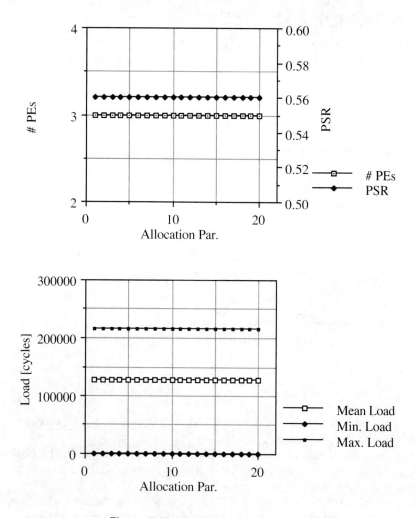

Figure 7.22 Allocation of the FFT12_1 FFT

Sun SPARCstation IPX). For this kind of algorithm, a different approach for partitioning must be chosen, as will be discussed in Chapter 8.

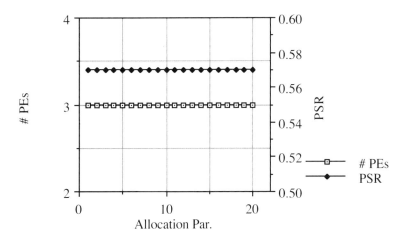

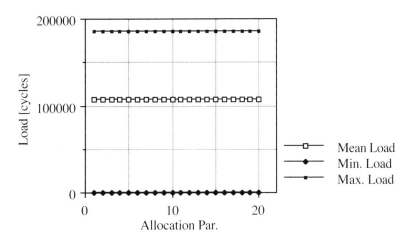

Figure 7.23 Allocation of the FFT12_2 FFT

Table 7.8 Partitioning Data of the FFTs

example	FFT12_1	FFT12_2	FFT12_3	FFT24_1	FFT24_2
serial time	382'362	322'974	321'798	1'412'228	n/a
# nodes	3	3	1408	4	~4'000
partitioned exec. time	384'256	324'868	n/a	1'425'290	n/a
# tasks	3	3	n/a	4	n/a
max. # links used	4	4	n/a	4	n/a

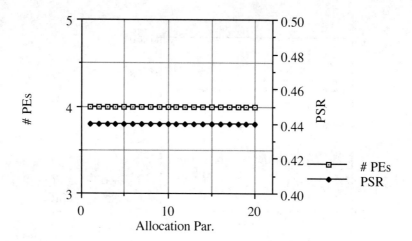

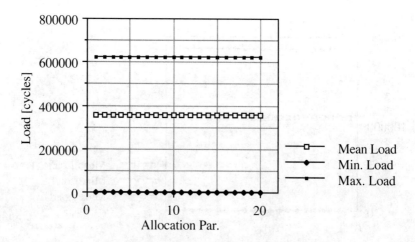

Figure 7.24 Allocation of the FFT24 FFT

For those examples for which usable results could be found, the corresponding values are displayed in the following figures.

For the 12-point FFT with no function expansion (FFT12_1), the results shown in Figure 7.22 illustrate the futile attempts to parallelize the computations. The main computations consist only of the call statements of the two stages of the transform and of the constant coefficients. Even if three PEs are utilized, no speed-up will result since the two stages of the FFT have to be computed sequentially.

For the 12-point FFT with expansion of the small index computation functions (FFT12_2) of Figure 7.23, the main deficiency remains the same as in Figure 7.22. The two stages of the FFT keep their sequential order so that no speed-up is achievable.

For the 24-point FFT of Figure 7.24 the same conclusions have to be drawn as for the preceding examples of the 12-point FFTs, that is, that the inherent sequential order of the stages remains.

As long as it is not possible to expand fully the stages of the transform, no satisfactory parallelization of FFTs can be expected.

7.7 Conclusions

The experiments conducted with the examples of all the kinds of algorithms targeted show that it is possible to generate automatically a parallel implementation of the computations. The only exception so far is the fast Fourier transform, where the high computation times prohibit the full expansion of the computations. This expansion is a prerequisite for finding a good parallel allocation of the algorithms.

As the examples with the LQR controller with estimator show, structural properties of the algorithms are exploited. The application algorithm is divided into blocks of equal size if the computations consist of connected blocks, e.g., of several scalar products. All the same, computations without this regular kind of structure are well distributed if a certain number of tasks is present, as the examples of the Runge-Kutta numerical integration method indicate. In the example given, the favourable effect of function expansion is also demonstrated. By replacing the call of a function by the function itself, much overhead caused by the call and by communication is eliminated.

It is interesting to note that all eight PEs available in the system are used in none of the examples. Five or six PEs are sufficient, and for all examples configurations of as few as two PEs are found.

The best results with the smallest PSR (parallel-to-serial execution time ratio) as low as 0.33 work with two to four PEs, whereas typical values of the PSR are around 0.64, but always less than 0.9. It seems that the nature of the application algorithms and the kind of execution (i.e., non-overlapping computations of the single iterations) impose these bounds on the performance. Hence, it would be useless to build multiprocessor systems with dozens of PEs for the purpose of real-time signal processing. However, if multi-rate processing is used, or if some kind of hierarchical information processing such as on-line system identification and adaptation is superimposed on the basic algorithms, the utilisation of more PEs can be useful.

An additional basis for this conclusion is the fact that the tasks formed initially from the data flow graphs need so many interconnections that the solution which saves the most communication cost is the one keeping the tasks as close to each other as possible, i.e., on the same PE.

The fact that the interconnections for signal processing applications are so numerous explains the need for using links among the PEs for several edges of the data flow graph by link fusion or link multiplexing. However, the communication costs which are already quite high would only increase if message routing through intermediate PEs were introduced.

The allocation parameter signifies an average number of tasks per PE. Until there are fewer tasks on the PEs, the optimization goal is strictly to preserve parallelism without regard to constraints imposed by the target system.

The experiments show that the number of PEs used decreases with increasing values of the allocation parameter. The higher the allocation parameter chosen, the later the optimization switches to the goal of finding an allocation which is realizable on the particular target system. Obviously, the longer the optimization runs without respect to the limitations of the real system, the harder it becomes in the second phase of the allocation to find a solution meeting the communication link constraint while utilizing many PEs.

The optimum solution with the shortest estimated execution time and the best PE utilization lies in the range of $\frac{\#\,\text{Tasks}}{4} \ldots \frac{\#\,\text{Tasks}}{8}$. For parameter values much above this range, a solution with only one or two PEs is found in most cases, sometimes running more slowly than the serial version of the algorithm.

The reference serial time for computing the PSR is somewhat optimistic, however, since it is just the sum of all node execution costs, neglecting any communication overhead. For critical applications, only test runs of the programs on the real hardware can give the final answer to the question as to whether the real-time constraints are met.

As expected, the granularity of the task graph is small, with an average number of nodes per task of little more than two.

CHAPTER 8
Conclusions

8.1 Assessment of the Data Flow Approach for Real-time Systems

The results presented in Chapter 7 show that data flow graphs are a very suitable representation of computations to be implemented in parallel. It would be much harder, if not impossible, to gain the same insight into the properties of the application algorithms from any other model.

From a data flow graph, the entire parallelism in an algorithm can be identified. Through the two-phase static allocation algorithm described in Chapter 6, as much parallelism as possible is preserved. Towards the end of the allocation process, though, some parallelism has to be sacrificed to achieve a process configuration which is realizable on the given hardware. The computing system is used optimally due to the fact that the structure of the interconnections among the processors is fixed only during the allocation. Therefore, the communication resources are used where they are needed most.

Since it is possible to allocate the target algorithms statically, the high run-time overhead of pure data flow systems for task administration can be avoided. The efficient execution of the tasks synchronized by the availability of data is supported on the Transputer by the autonomous link controllers and the fast micro-coded scheduler.

All these factors contribute to an efficient realization of the application algorithms on multiple processors.

8.2 Future Work

As mentioned in Chapter 7, is it difficult to distribute fast Fourier transform algorithms with the current system. This is due not to any fundamental weakness of the partitioning and allocation process, but rather to the fact that these algorithms contain a large number of nodes. As soon as the number of nodes becomes thousands, the computation time for partitioning rises very rapidly and exceeds values feasible for practical purposes.

One possibility for building the tasks faster and reducing their number may be to sacrifice some parallelism already in the initial graph partitioning phase and to form some kind of super-clusters. Each of these super-clusters could contain several of the linear clusters currently used, together with a common origin or destination of the respective data paths. For example, a super-cluster could contain several branches of a tree reduction graph.

Ideally, the computational load of such a super-cluster would be a fixed fraction of the total load of the whole application, e.g., one hundredth of the load. A fixed maximum number of tasks would then always be formed for any size of the application algorithm.

This modification, however, would render it impossible to simulate the exact execution of the data flow graph during the partitioning phase, as done until now. While some approximation can be used for the determination of the execution time, special care must be taken for determining the serial execution order of the inputs of the tasks. This order is critical because of the threat of deadlocks.

A possibility for the reduction of communication and thus of the execution time is given in the way state information is handled. In the current system, all internal information about the state of a set of computations must be returned as output to the data acquisition firmware at the end of one computation cycle if it is to be re-used in the next cycle. It is then returned unchanged to the process in the next cycle. This philosophy causes unnecessary communication, but was adopted in order to have only "memoryless" processes. Keeping relevant information at the place where it is needed again in the next computational cycle could save unnecessary data transfers.

Overhead could be reduced significantly if the tasks allocated to the processing elements were re-serialized. Currently, all tasks are placed for parallel execution, and the run-time scheduler deals with running them. Since only one task can be active at a time on one processor, the quasi-parallel execution degenerates to some kind of serial execution. Therefore, one could eliminate the task administration overhead and create a predefined sequence of the tasks if the time of the availability of the input and output data is known for each task. However, this will be the crucial point of the matter. If the tasks are all connected and linearly dependent, it will be easy to find a serial order for them. But if unconnected tasks are allocated on the same PE, it will be more difficult to determine an order of the tasks which neither unnecessarily delays nor blocks the execution altogether. Nonetheless, the potential benefits of such an extension are large.

Finally the introduction of a more flexible data type system is mentioned, which will in fact eliminate some restrictions currently imposed on the use of SISAL. The use of the data type "record" will then mainly be allowed. However, it is currently not known when the draft of the revised OCCAM91 language will be released for use.

Of course, computing systems other than T800 Transputers can be programmed using this method. Once the node execution cost tables and the communication cost models have been established, T9000 Transputers or signal processors such as the TMS320C40 can be used as well.

The T9000 Transputer is more powerful than its predecessor. However, since virtual channels are used the communication cost model will be more difficult to determine.

There the communication cost becomes dependent on the topology of the communication network, or communication once again has to be restricted to directly connected PEs. However, the virtual channels allow a direct interconnection among all processors. It is therefore no longer possible for some PEs to lie idle because they cannot be reached by direct links even if there is enough parallelism in the application.

The TMS320C40 signal processor possesses six serial links to build a communication network. This is a clear advantage over the T800 Transputer, since the limited number of links leads to the greatest loss in exploitable parallelism in the allocation step. Due to its highly pipelined internal structure it will be relatively difficult to obtain reliable and reproducible figures for the node execution cost table for the TMS320C40.

It is possible to use a variety of hardware platforms for this kind of parallel processing. Most important is that communication is fast and deterministic in the sense that a message of a given size always takes the same amount of time to transmit. If communication cannot be modelled exactly, building tasks and allocating them to the PEs in the way described in this book may lead to poor performance of the system.

An increased interest in exploiting fine-grain parallelism for general-purpose computations now exists, as documented by the development of the Sparcle chip at MIT ([AgKuKr 93]). This chip is based on the Sparc chip by Sun Microsystems and includes additional hardware to tolerate memory and communication latencies to support fine-grain synchronization, and mechanisms to initiate communication actions to remote processors and to respond rapidly to synchronization events and message arrivals. While this chip has not been designed specifically for real-time applications, it clearly aims at efficient message-passing multiprocessing on a large scale.

OCCAM need not necessarily be the language of the target system. While it offers inherent constructs for communication, other features such as the clumsy and very limited type system impose restrictions on the formulation of the algorithms to be processed. Other languages such as C or even C++ offer far better features together with highly optimized compilers adapted to the specific target system. The message-passing mechanisms then have to be hidden in routines which are called instead of the send and receive commands of OCCAM.

8.3 Reference

[AgKuKr 93] A. Agarwal, J. Kubiatowicz, D. Kranz, B.-H. Lim, D. Yeung, G. D'Souza and M. Parkin, "Sparcle: An Evolutionary Processor Design for Large-Scale Multiprocessors," *IEEE Micro*, vol. 13, no. 3, pp. 48-61, 1993.

APPENDIX A

Specification of the Multicomputer

This appendix provides some additional data on the multiprocessor system built for this project and its associated high-speed data acquisition system.

The multi-Transputer system consists of processor boards available off the shelf, whereas the whole data acquisition hardware was designed and built in order to meet the special requirements of fast real-time data processing.

A.1 Multi-Transputer System

The IMS T800 Transputer ([Whitby 85], [INMOS 89]) is a 32-bit CMOS microcomputer with a 64-bit floating point unit. It has 4 kbytes on-chip RAM for high-speed processing and four serial communication links. The instruction set supports the efficient implementation of high-level languages and provides direct support for the OCCAM model of concurrency when either a single Transputer or a network of such chips are used. Procedure call, process switching, and typical interrupt latency times are in the sub-microsecond range. This property makes the processor very well suited for real-time systems.

The T800 consists of the following basic blocks, all interconnected by a 32-bit wide internal bus:

- 32 bit processor
- floating point unit
- system services
- timers
- 4 kbytes of on-chip RAM
- external memory interface
- event logic
- four serial link interfaces

The processor speed is set to 20 MHz, achieving a peak instruction throughput of 20

MIPS (Million Instructions Per Second) and of 10 MIPS sustained.

The IMS T800 provides high-performance arithmetic and floating point operations. The 64-bit floating point unit provides single- and double-length operations according to the ANSI-IEEE 754-1985 standard for floating point arithmetic. It is able to perform floating point operations concurrently with the processor, sustaining a rate of 2.2 Mflops (Million Floating Point Operations Per Second) at a processor speed of 20 MHz.

The standard INMOS communication links allow networks of Transputer family products to be constructed by direct point-to-point connections with no external logic. The IMS T800 links are used at a speed of 20 MBit/sec, but can also operate at 5 or 10 MBit/sec. Each link can transfer data bidirectionally at up to 2.35 MByte/sec.

All boards of the MultiCluster-1 series computer are manufactured by Parsytec Computer GmbH, Aachen, Germany (in the U.S.: Parsytec Inc., West Chicago).

The system consists of four MTM-2-11 boards with two Transputers each, running at a speed of 20 MHz. Each processor is equipped with two Mbytes of local RAM. The interconnection links are easily configurable by plug-in cables. An MTM-Mac board with two IMS T800 processors in a Macintosh II computer is used for compiling the programs and as a console for controlling the application programs.

An additional GDS (Graphic Display System) board allows the display of colour graphics on a separate 19 inch monitor. One VTF (Versatile Transputer Frontend) board with one T800 and two T222 (16-bit integer Transputer) processors provide the connections to the data acquisition system through two 16-bit wide data and address buses.

A.2 High-speed Data Acquisition System

The data acquisition system is designed for easy configuration to a variety of possible settings. Its structure is shown in Figure A.1 (adapted from [SchZan 90]). During start-up, the desired system configuration is read from a file and is set accordingly.

At most, 16 analog input signals can be captured. Each channel is equipped with a programmable high-accuracy instrumentation amplifier (with gains 1, 10, 100, 200, 500) and a switched-capacitor (SC) anti-aliasing filter of order four or eight. The anti-aliasing filter has a Bessel characteristic which makes it easy to model it as phase-shifting device due to its linear phase response. Filters of the Chebyshev type show a steeper descent in the amplitude response, but their nonlinear phase response makes it impossible to model the filter as a simple time delay for the controller design.

Since the switched-capacitor filter is a discrete-time system, it requires an anti-aliasing filter for itself, but with a cutoff frequency which lies far above the frequencies of interest. Therefore, this pre-filter is realized as a tunable active filter of second order which does not influence the input signals. The whole anti-aliasing filtering section may be bypassed if sampling unfiltered signals is desired.

The sampling rate is selectable anywhere in the range of 100 samples/sec to 67 kSamples/sec. In the uppermost range above 33 kSamples/sec, only four analog signals can be sampled. The settings for the anti-aliasing filters are computed automatically in

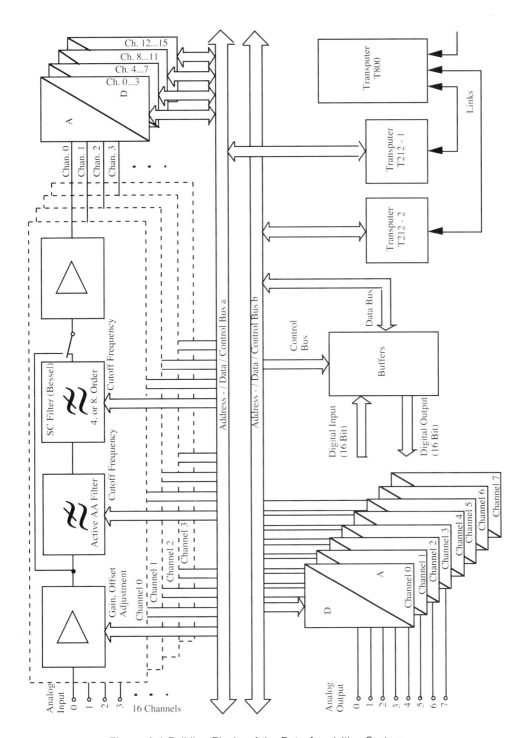

Figure A.1 Building Blocks of the Data Acquisition System

152   Specification of the multicomputer

accordance with the sampling rate.

The resolution of the analog-to-digital converters is 12 bits, as for the eight digital-to-analog converters. Digital input and output is possible through a buffer of 16 bits width.

For the control of a six-cylinder internal combustion engine, a supplementary board containing six custom chips for the individual control of the injection and ignition (see [Geerin 89]) was developed (not shown in Figure A.1).

A.3 Firmware

The firmware belonging to the data acquisition system is used in two areas:

- setting up the system
- delivering data to and receiving data from the application program

As described in Section A.2, the desired configuration (number of channels sampled, sampling rate, order of anti-aliasing filter) is read from a file. The single hardware components are then programmed accordingly. This is done by a program running on a Transputer located in the Macintosh II which is used as file server and terminal. From this console, processing is started and stopped.

Once processing has started, the firmware reads the input values from the A/D converters as soon as they are available. Since the converters are clocked from a hardware timer, processing is event-driven. After adjusting the offset voltage by software and scaling the values, data are transmitted through links to the application program running on the eight T800 processors. The results computed by the application program are returned to the digital-to-analog converters.

A sophisticated timing control is active during the processing. It checks the times when input data are sampled and when the corresponding output values are returned to the D/A converters. If the difference between these times is larger than the sampling period, an error message is sent to the console and processing stops. Then either the sampling period is too short or the processing time is too long and must be reduced, e.g., by distributing the computations to the processors in a different way.

A.4 References

[Geerin 89] H.P. Geering, "ICX: A custom VLSI-chip for engine control," *Int. J. of Vehicle Design*, vol. 10, no. 5, pp. 592-597, 1989.

[INMOS 89] INMOS Ltd., *The Transputer Databook*, Second Edition 1989.

[SchZan 90] D. Schweizer and R. Zanetti, "Hochgeschwindigkeits-Datenerfassungssystem für Multitransputer-Rechner," Diploma Thesis in Electrical Engineering, Measurement and Control Laboratory, Swiss Federal Institute of Technology (ETH), Zurich, February 1990.

[Whitby 85] C. Whitby-Strevens, "The Transputer," in *Proc. 12th Ann. Symposium Comp. Architecture*, 1985, pp. 292-300.

APPENDIX B

Detailed Specification of the Processing Steps for Partitioning and Allocating

In this appendix the processing steps for partitioning and allocating the data flow graphs are outlined in more detail than is possible in the main text. Data used for some of the processing stages, e.g., node execution costs, are presented in tables.

The algorithms are formulated in pseudo-code related to the syntax of the Pascal programming language.

The actual programs for performing the operations described are written in Pascal and consist of around 18 000 lines of code, not counting the SISAL compiler.

The relations between the sections of Appendix B and the text are as follows:

- Section B.1   Communication Analysis   → Section 4.2
  - Section B.2   Graph Expansion   → Section 4.3
  - Section B.3   Execution Cost Analysis   → Section 4.4
  - Section B.4   Graph Partitioning   → Chapter 5
  - Section B.5   Task Allocation   → Chapter 6
  - Section B.6   OCCAM Code Generation   → Section 6.6

B.1 Communication Analysis

The amount of data transferred over the edges of a graph gr is analysed by starting with the graph's output edges. For each edge inspected, the following procedure is applied. If it carries a data item with a basic data type then the item's size is read from Table B.1. If the edge carries a data item with a compound data type, then the preceding node and its input edges are inspected in order to determine the data type's size. If this is not possible, the graph is traversed recursively until all necessary information has been found. Then all inputs of the edge's source node are inspected in the same manner.

Processing continues until all edges of the graph have been inspected.

The structure of the procedure for the communication analysis is shown below.

Table B.1 *Size of the Data Types*

SISAL Type	Size (Bytes)	Corresp. OCCAM Type
Boolean	1	BOOL
Character	1	BYTE
Integer	4	INT
Real	4	REAL32
Double_Real	8	REAL64
Array[Component Type]	# elements × component size	[# elem.]Component Type

<u>COMMUNICATION</u>(fgr):
for all output edges edg of fgr **do**
   <u>determine_data_type</u>(edg)
   if edg is no input edge of fgr **and**
     source node snd has not been visited yet **then**
     <u>inspect_node</u>(snd, edg)
   end if
end for all
end communication

determine_data_type(edge):
case data type of **of**
   basic: read data size in size table
      set label edge.comm
   compound: find source node snd of edge
      if this is not possible since edge is an input edge of
        the function graph **then**
        determine the value number_of_elements of the
          constant input beside edge
        set edge.comm := number_of_elements ×
          size of basic type of edge
     else
        <u>inspect_node</u>(snd, edge)
        set label edge.comm to size returned by
          inspect_node
     end if
end case
end determine_data_type

```
inspect_node(nd, edge):
case nd of
   simple node:    determine the size of the data type carried by
                      edge according the function of nd
                   if necessary determine the size of
                      nd's input edges
                   return the size of edge's data type
   compound node: inspect all subgraphs of nd
                   determine the size of the data type carried by
                      edge according the function of nd
                   if necessary determine the size of
                      nd's input edges
                   return the size of edge's data type
end case
for all input edges edg of nd do
   determine_data_type(edg)
   if edg is no input edge of the function graph and
      source node snd has not been visited yet and
      edg has not been visited yet then
         inspect_node(snd, edg)
   end if
end for all
end inspect_node
```

B.2 Graph Expansion

The graph *graph* and its associated function graphs are visited in a bottom-up, depth-first way. Starting with the output edges' source nodes, all preceding nodes are inspected until the top of the graph is reached.

The following functions are performed:

- NoOp nodes are eliminated
- Two consecutive Not nodes are eliminated
- All compound nodes except Forall are left unchanged
- Forall nodes are replaced by the same number of copies of the body graph as there are iterations in the loop. The generator subgraph is replaced by a node which generates all indices for the body graphs or distributes the elements of a vector to the body graphs. The returns subgraph is replaced by the appropriate reduction graph connected to the body graphs.
- If the range generator node has only constant inputs, the node is removed and the constant indices are fed directly to the body graphs.

- The parameter "*nodecount*" sets a minimum threshold for the number of nodes of functions called. The call node is replaced by the function graph only if the function contains more nodes than *nodecount*. The aim of this threshold is to prevent the graph from expanding too quickly by the insertion of large functions.

For replacing the "Reduce" nodes which collect the results of all loop iterations in Forall nodes, a structure is built from nodes according to Table B.2.

Table B.2 *Replacement of "Reduce" Nodes*

Reduction	Function	Data Type	Node
Sum	+	Int, Real, Double	Plus
	OR	Boolean	Plus
Product	×	Int, Real, Double	Times
	AND	Boolean	Times
Least	minimum	Int, Real, Double	Min
	AND	Boolean	Min
Greatest	maximum	Int, Real, Double	Max
	OR	Boolean	Max
Catenate	catenation	array(T)	ACatenateExp

In addition to the nodes defined in the IF1 manual ([SkeGla 85], see Section 4.5), the new nodes shown in Table B.3 are used. They are inserted during the expansion of a Forall node when the generator subgraph is eliminated and for the substitution of the returns subgraph. When the values to be collected are accompanied by Boolean values it cannot be decided at compile-time which values are to be discarded. Hence it is impossible to build a static reduction structure. The reduction nodes AGatherExp, FirstValueExp, FinalValueExp, and RedLeftExp are therefore defined. These decide at run-time which values of the loop bodies are to be utilized for computing the final result.

In the node cost Table B.4 additional nodes appear which have been introduced for the distinction of calls to different intrinsic functions and for different reduction functions. Their functions are explained by their names, e.g., Call_Sin for a call to the sine function.

Array(T) means an array with elements of type "T". For the description of the node function, the same notation as in the IF1 manual is used.

Graph expansion

Table B.3 *Additional Graph Nodes*

Node Name	No.	Function Description
AGatherExp	200	integer × {T × Boolean} → array(T) Creates an array with the first element at the position indicated by the integer input. The value of an input is inserted in the array if its associated Boolean input carries the value "True".
AScatterExp	201	array(T) → {T × integer} Possesses for each element of the input array a pair of output edges with the array element and its index. Each pair leads to one body graph.
RangeGenerateExp	202	integer × integer → {integer} Same function as RangeGenerate except that the output edge with multiple integer values is replaced by multiple integer outputs. Each output leads to one body graph.
FirstValueExp	203	{T × Boolean} → T Performs run-time check of the Boolean inputs. The lowest input whose associated Boolean input is "True" is forwarded to the output.
FinalValueExp	204	{T × Boolean} → T Performs run-time check of the Boolean inputs. The highest input whose associated Boolean input is "True" is forwarded to the output.
RedLeftExp	205	function × T × {T × Boolean) → T Every input for which the associated Boolean input is "True" is reduced according the function specified.

GRAPHEXP(graph, nodecount):
insert all top level graphs of graph in the list "functiongraphs"
for all toplevel graphs gra **do**
    for all output edges e of gra **do**
        find predecessor node prednode
        <u>inspect predecessor</u>(prednode)
    end for all
    for all ABuild nodes anode
        which have not yet been visited **do**
        if anode has only literal inputs **then**

158 Detailed specification of the processing steps

```
            expand_abuild(anode)
         end if
      end for all
   end for all
end graphexp

inspect predecessor(pred)
case pred.type of
      compound node   :      inspect compound(pred)
      simple node     :      inspect simple(pred)
end case
for all input edges of pred do
   find predecessor node pnode in pred's graph
   if found
      inspect predecessor(pnode)
   end if
end for all
end inspect predecessor

inspect simple(nd):
case nd of
   Call : case function called of
         sqr      :    replace nd by Times node,
                       duplicate input
         controlalgorithm : do nothing
         otherwise :      find function graph fgr
            if found and # of nodes<nodecount then
                  graphexp(fgr, nodecount)
                  connect input and outputs of nd
                       to fgr
                  remove graph boundary of fgr
                  eliminate nd
                  nd := source node of old input
                       edge to fgr
                  inspect_pred(nd)
            end if
         end case
   NoOp : eliminate nd and rewire the edges
   Not  : find predecessor node pnd in the nd's graph
         if pnd is Not then
            connect input edge of pnd to output edge of nd
            eliminate nd and pnd
         end if
   otherwise: do nothing
```

end case
end inspect simple

inspect compound(nd):
case nd **of**
   Forall   :   expand Forall(nd)
   Select   : **for all** alternative graphs agr **do**
          graphexp(agr, nodecount)
       end for all
   Iter   : graphexp(graph of body, nodecount)
   LoopA, LoopB : **with** the four subgraphs
          initialization, test, body, returns **do**
          graphexp(subgraph, nodecount)
       end with
   TagCase :   error(illegal operation on (nd.srcline))
end case
end inspect compound

expand Forall(fnode):
ggr := generator subgraph
bgr := body subgraph
rgr := results subgraph
for all nodes gnode of ggr **do**   (* inspect GENERATOR *)
   case gnode **of**
      RangeGen   :   create new gnode_1 RangeGenExp
      AScatter   :   create new gnode_1 AScatterExp
   end case
   connect the input edges of gnode_1 to the corresponding
     edges leading to fnode
   set communication and data type of the edges
end for all
remove graph boundary of ggr
graphexp(bgr, nodecount)   (* inspect BODY *)
for i = 0 ..fnode.range **do**
   duplicate_graph(bgr, bodylist[i])
   connect the inputs of the nodes in bodylist[i] to the outputs
     of the generator node gnode_1
     or to the sources outside fnode of the direct inputs
   set communication and data type of edges
   remove graph boundary of bodylist[i]
end for
for all output edges oedge of rgr **do**   (* inspect RESULTS *)
   find the source node snode of oedge
   visit reduce node(snode, bodylist)

```
      end for all
      connect outputs of rgr to output edges of fnode
      remove graph boundary of rgr
      for all nodes gnode of expanded Forall node do
        if gnode.name=RangeGenExp
           and gnode has only literal inputs then
              replace all the outputs by literal inputs with appropriate value
              delete node gnode
        end if
      end for all
      delete node fnode
   end expand Forall
```

B.3 Execution Cost Analysis

All the graphs are traversed in a linear way. The edges are assigned an execution cost according the cost model developed in Section 4.4, assuming that they are realized as external links. The nodes' cost is read from Table B.4 which contains their execution time depending on the data type of the arguments. The cost of some nodes also depends on the number of inputs.

For the unexpanded loop nodes (LoopA, LoopB, Iter), a default number of iterations (the so-called range of the node) is assumed which is read from a parameter file first. Thus, this default value can be changed by the user. In the Select node used to represent if-statements, the graphs for the alternatives are inspected and their execution cost is determined. The Select node itself is assigned the maximum cost of any of its alternative graphs.

Table B.4 *Node Execution Cost Table*

Node	No.	Char	Bool	Integer	Real	Double
Forall	0	not av.				
Select	1	$9+4\times A$	$9+4\times A$	$9+4\times A$	$9+4\times A$	$9+4\times A$
TagCase	2	n/a				
LoopA	3	9	9	9	9	9
LoopB	4	9	9	9	9	9
Iter	6	23	23	23	23	23
AAddH	100	$5+14\times Ie$	$5+15\times Ie$	$9+11\times Ie$	$9+11\times Ie$	$7+18\times Ie$
AAddL	101	$5+14\times Ie$	$5+15\times Ie$	$9+11\times Ie$	$9+11\times Ie$	$7+18\times Ie$
AExtract	102	$5+14\times Oe$	$5+15\times Oe$	$9+11\times Oe$	$9+11\times Oe$	$7+18\times Oe$
ABuild	103	$8+15\times I$	$8+15\times I$	$8+11\times I$	$8+11\times I$	$8+18\times I$
ACatenate	104	$5+14\times Oe$	$5+15\times Oe$	$9+11\times Oe$	$9+11\times Oe$	$7+18\times Oe$

Table B.4 (*continued*)

AElement	105	21	21	20	21	27
AFill	106	25×*I*	20×*I*	28×*I*	33×*I*	41×*I*
AGather	107	37	41	41	41	42
AIsEmpty	108	9	9	9	9	9
ALimH	109	3	3	3	3	3
ALimL	110	3	3	3	3	3
ARemH	111	5+14×*Oe*	5+15×*Oe*	9+11×*Oe*	9+11×*O*	7+18×*Oe*
ARemL	112	5+14×*Oe*	5+15×*Oe*	9+11×*Oe*	9+11×*Oe*	7+18×*Oe*
AReplace	113	5+14×*Oe*	5+15×*Oe*	9+11×*Oe*	9+11×*Oe*	7+18×*Oe*
AScatter	114	n/a				
ASetL	115	5+14×*Oe*	5+15×*Oe*	9+11×*Oe*	9+11×*Oe*	7+18×*Oe*
ASize	116	3	3	3	3	3
Abs	117	0	0	18	35	48
BindArg.	118	n/a				
Bool	119	0	0	6	0	0
Call	120					
sin	10	0	0	312	296	486
cos	11	0	0	278	262	322
tan	12	0	0	376	360	395
asin	13	0	0	241	225	337
acos	14	0	0	250	234	346
atan	15	0	0	314	298	288
sqrt	16	0	0	139	123	253
sqr	n/a	0	0	39	23	37
log	17	0	0	517	501	905
ln	18	0	0	106	90	482
log10	19	0	0	142	126	820
etothe	20	0	0	470	454	717
rand	21	0	0	232	216	414
Char	121	0	0	32	0	0
Div	122	0	0	51	31	48
Double	123	0	0	0	11	0
Equal	124	12	12	12	21	24
Exp	125	0	0	0	417	1112
FirstValue	126	13	13	16	16	20
w Bool Inp	22	19	19	22	22	26
FinalValue	127	0	0	0	0	0
w Bool Inp	23	11	11	14	14	18
Floor	128	0	0	0	17	18

Table B.4 (*continued*)

Int	129	4	4	0	17	18
IsError	130	n/a				
Less	131	11	11	11	22	25
LessEqual	132	13	13	13	24	27
Max	133	0	14	19	33	37
Min	134	0	14	19	33	37
Minus	135	0	0	11	19	22
Mod	136	0	0	48	53	58
Neg	137	0	0	7	17	19
NoOp	138	n/a				
Not	139	0	6	0	0	0
NotEqual	140	14	14	14	23	26
Plus	141	0	14	11	19	22
RangeGen.	142	n/a				
RBuild	143	n/a				
RElements	144	n/a				
RReplace	145	n/a				
RedLeft	146					
Sum	24	0	14	11	19	22
w Bool inp	25	0	19	15	23	29
Product	26	0	14	29	29	47
w Bool inp	27	0	18	36	46	52
Least	28	0	14	19	33	37
w Bool inp	29	0	16	17	30	36
Greatest	30	0	14	19	33	37
w Bool inp	31	0	19	17	30	36
Catenate	32	0	15+15×I	19+11×I	19+11×I	17+18×I
w Bool inp	33	0	20+15×I	24+11×I	24+11×I	21+18×I
RedRight	147	n/a				
RedTree	148	n/a				
Reduce	149	n/a				
AllB.LastV	150	3	3	7	6	10
w Bool inp	34	9	9	13	12	16
Single	151	0	0	14	0	14
Times	152	0	11	23	23	37
Trunc	153	0	0	0	20	19
AGath.Ex	200	8+22×I	8+22×I	8+24×I	8+23×I	8+25×I
AScat.Exp	201	3×O	3×O	7×O	6×O	10×O
R.G.Exp	202	0	0	0	0	0

Execution cost analysis   163

Table B.4 *(continued)*

FirstV.Exp	203	90	90	93	93	97
Fin.V.Exp	204	8×E +15×I	8×E +15×I	9×E +15×I	10×E +15×I	12×E +15×I
ARedL.E.	205					
Sum	35	0	3+32×I	4+29×I	5+34×I	7+38×I
Product	36	0	3+26×I	4+31×I	5+38×I	7+51×I
Least	37	0	3+26×I	4+29×I	5+32×I	7+35×I
Greatest	38	0	3+32×I	4+29×I	5+32×I	7+35×I
Catenate	39	18+29×I	18+30×I	18+26×I	18+26×I	18+33×I

The meaning of the letters used in Table B.4 is:

A	number of alternative graphs in the Select node
E	number of elements in an array
I	number of input edges
Ie	number of array elements of the input edge
O	number of output edges
O	number of array elements of the output edge
n/a	not available

The nodes with the numbers 10-39 and 200-205 do not appertain to the original IF1 standard, but have been defined for this project.

All durations are expressed in terms of instruction cycles. One cycle of the T800 Transputer lasts 50 nanoseconds at a clock frequency of 20 MHz.

The structure of the cost assignment routine is shown below.

```
COSTASSIGN(G):
read parameterfile with default values for loop nodes
for all graphs gra included in G do
    graphcost(gra)
end for all
end costassign

graphcost(gra):
sum := 0
for all nodes nd of gra do
    nd.costs := nodecost(nd)
    sum := sum + nd.costs
    for all input edges e_in of nd do
```

```
        e_in.commkind := external
        e_in.costs := edgecost(e_in)
        sum := sum + e_in.costs
    end for all
end for all
for all output edges e_out of gra do
    e_out.commkind := external
    e_out.costs := edgecost(e_out)
    sum := sum + e_out.costs
end for all
gra.costs := sum
end graphcost
```

edgecost(edg):
```
words := (edg.comm + 3) div 4  (* number of words read *)
case edg.commkind of
    external :  edgecost := words × T_meme + edg.comm × T_trans
    internal :  edgecost := words × T_memi
    variable :  edgecost := 0
end case
end edgecost
```

Rendered with math:

$$\text{words} := (\text{edg.comm} + 3) \text{ div } 4$$

external: $\text{edgecost} := \text{words} \times T_{meme} + \text{edg.comm} \times T_{trans}$
internal: $\text{edgecost} := \text{words} \times T_{memi}$
variable: $\text{edgecost} := 0$

nodecost(nd):
```
case node type of
    simple  : if nd.name=Call then
                case function called of
                    sin     :   nd.name := Callsin
                    cos     :   nd.name := Callcos
                    tan     :   nd.name := Calltan
                    asin    :   nd.name := Callasin
                    acos    :   nd.name := Callacos
                    atan    :   nd.name := Callatan
                    sqrt    :   nd.name := Callsqrt
                    log     :   nd.name := Calllog
                    ln      :   nd.name := Callln
                    log10   :   nd.name := Calllog10
                    etothe  :   nd.name := Calletothe
                    rand    :   nd.name := Callrand
                otherwise   begin
                                find function graph cgr
                                graphcost(cgr)
                                nd.costs := cgr.costs
                            end
                end case
```

        else
           determine data type dt of output edge
           determine node costnd.costs from cost table
        end if

  compound :   **case** nd **of**
      Forall   :   error(illegal opration on (nd.srcline))
      Select :   <u>graphcost</u>(selector graph)
          for all alternative graphs agr of nd **do**
              set for all edges of agr
              communication kind := variable
              <u>graphcost</u>(sgr)
          end for all
          sum := max(agr.costs)
          sum := sum + overhead cost
          determine number nalt of alternatives
          sum := sum + nalt × cost per alternative
          nd.costs := sum
      TagCase : error(illegal operation on (nd.srcline))
      LoopA :   **for all** subgraphs sgr of nd **do**
              set for all edges of sgr
              communication kind := variable
              <u>graphcost</u>(sgr)
          end for all
          sum := (body.costs + test.costs +
                returns.costs) × nd.range
          sum := sum + initialization.costs
          if nd = LoopB **then**
              sum := sum + test.costs
          end if
          nd.costs := sum + overhead cost in table
      Iter   :   set for all edges of body graph
          communication kind := variable
          <u>graphcost</u>(body)
          nd.costs := body.costs × nd.range +
              overhead cost from table
      end case
end case
nodecost := nd.cost
end nodecost

B.4 Graph Partitioning

An annotated data flow graph *fgr* is partitioned into tasks as described in Chapter 5. Initially, all nodes represent a task of their own with only external communication links among them. Then, the tasks are temporarily merged pairwise and the pair yielding the lowest execution time is joined definitively. This search is continued until the execution time has reached its minimum. The execution time is determined by simulating the execution of the dataflow graph with the routine "simulate".

The result of the routine "partition" is a list of tasks with the appertaining nodes. For each node, the order of the execution of the input and output statements is determined. This order ensures freedom from deadlock (output statements) and minimum idle time (input statements).

PARTITION (fgr):

prepare graph
sort_outputs(fgr)
mintime := simulate(fgr)
repeat
   for all nodes nd in fgr with only one external output oed **do**
      if sink node of oed has only external and constant inputs **then**
         join sink node of oed and nd in one process
         replace oed by variable
         sort_outputs(fgr)
         executiontime := simulate(fgr)
         if executiontime<mintime **then**
            minedge := oed
            mintime := executiontime
         end if
         separate the two nodes again, restore original state
      end if
   end for all
   join nodes connected by minedge in one process
   replace minedge by variable
until execution time has reached the minimum
end partition

simulate(fgr):
insert input node and all nodes of fgr with only literal inputs in circular list
while list **not** empty **do**
   pick current node cnode from list
   node_inwork := true
   pick first output edge aout that has not yet been visited

Graph partitioning

```
if output not performed yet then
   while not all output edges have been visited
           and not node_inwork do
      if external edge then
         clock := clock + Tset-up
         ticks := ticks + Tset-up
      end if
      aout.outtime := clock
      if all inputs of successor node snode are ready then
         process successor(snode)
         if external edge then
            clock := aout.rendezvous
            ticks := ticks + Tset-up + commun. cost
         end if
         pick next outport aout
      else
         node_inwork := false
      end if
   end while
else (* output already performed *)
   if aout.rendezvous > 0 then
      (* rendezvous has taken place in the meantime *)
      clock := aout.rendezvous
      if external edge then
         ticks := ticks + Tset-up +  communication cost
      end if
      pick next outport aout
   else (* successor node not yet executed *)
      while not all outp. edg. have been visited
              and not node_inwork do
         if all inputs of succ. node snode are ready then
            process successor(snode)
            if external edge then
                clock := aout.rendezvous
                ticks := ticks + Tset-up + comm. cost
            end if
            pick next outport aout
            if external edge then
                clock := clock + Tset-up
                ticks := ticks + Tset-up
            end if
            aout.outtime := clock
         else
            node_inwork := false
```

```
            end if
          end while
        end if
      end if
      if node_inwork and all edges visited then
          remove cnode from list
      end if
      cnode := next node in list
end while
simulation time := time of clock of result node
end simulate

process successor(snode):
sort inports inp according times when input data available
pick first input ain
if there is a pred. node pnode of snode in the same process as snode
then
    while there are inputs ain which have not yet been visited do
        if pnode not executed yet then
            if external edge then
                freetime := clock+Tset-up-out time of corr. output
            else
                freetime := clock-out time of corr. output
            end if
            if freetime>pnode.cost then
                (* enough time to execute pnode *)
                clock := clock + pnode.cost
                pnode.vistited := true
                ticks := ticks + pnode.cost
            else
                ain.pre-emptive := true
            end if
        end if
        if external edge then
            ain.intime := clock + Tset-up
            clock := max(ain.intime, out time of corr. outp.) +
                    communication cost
            ticks:=ticks+Tset-up+ communication cost
        else
            ain.intime := clock
            clock := max(ain.intime, out time of corr. outp.)
        end if
        rendezvous := clock
        pick next input edge ain
```

```
      end while
      if not pnode visted then
         (* execute pnode before beginning with snode *)
         clock := clock+ pnode.cost
         pnode.visited := true
         ticks := ticks + pnode.cost
      end if
   else (* snode is the first node in process *)
      for all inputs ain do
         ain.intime := clock + Tset-up
         rendezvous := max(ain.intime, out time of corr. outp.) +
                  communication cost
         clock := rendezvous,  ticks := ticks+Tset-up+ain.cost
      end for all
   end if
   if snode last node in process then  (* execute function of snode *)
      clock := clock + snode.cost
      ticks := ticks + snode.cost
      snode.visited := true
      pick first external output aout (* mark first output *)
      clock := clock+ Tset-up
      ticks := ticks + Tset-up
      aout.outtime := clock
      if all inputs of successor node snd are ready then
         process successor(snd)
         clock := rendezvous,  ticks := ticks + Tset-up + aout.cost
         pick next outport aout
      end if
   else (* defer execution of snode *)
      snode.deferred := true
   end if
   insert snode in list
end process successor
```

B.5 Task Allocation

The partitioned data flow graph *fgr* is allocated on the target machine according to the rules outlined in Chapter 6. The aims are to reach a minimum execution time while respecting the limitations imposed by the real target machine. The target machine consists of *numberofpe* processors with *noflinks* serial communication links each. It is not specified how these links are interconnected. The structure of the interconnection network results only from the allocation procedure.

170 Detailed specification of the processing steps

The allocation begins by placing each task on a processing element (PE) of its own. Subsequently, in each iteration one PE is removed from the system by either moving all its tasks to another PE (phase one) or distributing the tasks to adjacent PEs (phase two). The iteration continues until there are not more PEs than are available in hardware (*excesspe*≤0) and until on any PE the number of links used is less than or equal to the number of links available in hardware (*excesslinks*≤0).

In phase one, all processes of a PE are moved to a connected PE and the PE is removed from the system. The PE for which the total number of links in the system is reduced most is selected for elimination. Switching from phase one to phase two is based on the parameter *alpha*. As soon as *av_tasknumber* (the average number of tasks per PE, initially one) reaches *alpha*, phase two is activated.

In phase two, that PE is chosen for elimination which minimizes the weighted sum of the links in the system. In this sum, for each PE the unused links are counted once. The links exceeding the number of the available links are counted fivefold. By this target function the elimination of surplus links is enforced. In order to distribute the load more evenly among the PEs, the task set of a PE is not moved as a whole to another PE as in phase one. Instead, each task is placed on the PE to which it communicates most.

The number of links is further reduced by additional optimizations for link fusion and link multiplexing. With link fusion, two links from the same source node to different sink nodes located on the same PE are fused. Data are transferred over only one link to the PE where the nodes are located and are then distributed to them. With link multiplexing, ordinary time multiplexing of a link is introduced. The source and sink nodes of the communication channels may belong to any task on the two PEs involved.

The result of the routine *allocation* is a list of the PEs and the tasks allocated to them. For each PE, another list is created which contains its links and their destinations.

<u>ALLOCATION</u> (fgr, alpha):

```
prepare allocation
links in system(maxlinks, totlinks)
excesspe := maxpe - numberofpe
excesslinks := maxlinks - numberoflinks_per_PE
while (excesspe>0) or (excesslinks>0) do
    if av_tasknumber<alpha then
        build list pairlist of all connected PE pairs sorted according
            their total communication cost
        maxdiff := 0
        mergepair := nil
        for all pairs PE1, PE2 of the first half of pairlist do
            determine no. of links used by both PEs before
                merging: links_before
            determine number of links used after merging:
                links_after
                    (* including link fusion and link multiplexing *)
```

```
              linkdiff := links_before-links_after
              if linkdiff>maxdiff then
                  maxdiff := linkdiff, mergepair := (PE1, PE2)
              end if
          end for all
          join_processes(mergepair)
          delete pairlist
      else (* av_tasknumber≥alpha *)
          minpe.totlinks := infinity
          for all processors PEi do
              distribute_processes(PEi, processlist)
              (* place tasks to the PE they are most connected to
                         (lowest level first) *)
              links_in_system(noflinks, total)
              if total<minpe.total then  (* keep PEi as minpe *)
                  minpe.pe1 := PEi
                  minpe.linksused := noflinks
                  minpe.totlinks := total
              end if
              collect_processes(PEi, processlist)
          end for all
          distribute_processes(minpe.pe1, processlist)
      end if
      links_in_system(maxlinks, total)
      excesspe := excesspe - 1
      av_tasknumber := numberoftasks / maxpe
      excesslinks := maxlinks - numberoflinks_per_PE
end while
end allocation


prepare allocation:

for all PEs ape do
    compute the cost of the processes =
        $\sum$ cost of all nodes in the process
    compute the load of ape = $\sum$ cost of all processes on ape
end for all
create_links(first PE, linkopt)
create adjaceny matrix tm of tasks
compute levels of task nodes = max. number of task nodes visited on all
    paths from data entry point to the task node investigated
end   prepare_allocation

join_processes(pe1, pe2):
```

change all channels between processes of pe2 and processes of pe1 to internal channels
append the list of processes of pe2 to the list of processes of pe1
delete links between pe1 and pe2
change all remaining links of pe2 to simple links
append them to the list of links of pe1
<u>optimize_links</u>(pe1, true)
remove pe2 from the system
end join_processes

<u>distribute_processes</u>(pe1, plist):

for all tasks pr on pe1 **do** in ascending level order
    find processor pe2≠pe1 pr is most connected to
        (according communication costs)
    remove pr from pe1 (including links)
    insert pr on pe2 (including links)
    increasepe2.load by the cost of pr
    insert pr in plist
end for all
for all PEs pe2 where links have been modified **do**
    <u>optimize_links</u>(pe2, true)
end for all
end distribute_processes

<u>collect_processes</u>(pe1, plist):

for all processes apr in plist **do**
    remove apr from processor ope
    insert apr on pe1
    increasepe1.load by the cost of pr
    decrease ope.load by the cost of pr
    remove unnecessary links from ope
    create links on pe1 and on then connected PEs
    for all PEs pe2 where links have been modified **do**
        <u>optimize_links</u>(pe2, true)
    end for all
end for all
end collect_processes

<u>links_in_system</u>(maxlinks, totlinkno):

returns for the whole system
    maxlinks   : max. number of links used on any PE in the system

```
    totlinkno    : total number of links used of all PEs in the system
for all PEs do
    determine the number of links lno
    if lno>noflinks then
        totlinkno := totlinkno + 5 (lno-noflinks)
    else
        totlinkno := totlinkno + (noflinks-lno)
    end if
end  links_in_system
```

optimize_links(fpe, only_one_pe):

Reduces the number of links of fpe (only_one_pe=true) or of all PEs by link fusion or link multiplexing, if possible.
Links between two PEs can be either fused or merged, but not both.
    link fusion: two links emanating from the same source node are
               fused in one.
        Data are duplicated and distributed to the sink nodes on
           the target PE.
    link multiplexing: common time-multiplexing of one physical link.
        Several logical channels from one PE to another are
           placed on the same link.
First all links are inspected for link fusion. Then all links which are not fused are inspected for link multiplexing.
Then for all links which are used only monodirectionally a matching counterpart is searched. These two monodirectional links between PE A and PE B are unified. Thus they are replaced by one bidirectional link.
end optimize_links

B.6 OCCAM Code Generation

Once the graph *fgr* has been partitioned and allocated, each node is positioned in a process. Each process is placed on a PE, and for each edge it has been determined how it is created (as variable, internal channel, or external channel).

Therefore, for each PE used in the system a declaration of the interprocessor channels is generated. These channels are placed on the physical links of the Transputers. Together with the headers of processes placed on each PE, this makes up the contents of the allocation file "*occam.all*."

The OCCAM source code of all the processes of the PEs is written to the file "*occam.src*." For each PE with number *i* one main process PEi is generated which handles the communication with the other PEs over the external links. The multiplexing and distributing processes for those links that carry several graph edges are placed there for parallel execution, together with all the other processes allocated on the PE. For these links, the protocol definitions are included. The so-called fused edges are realized by tagged communication, with tags labelled Tk where k is a consecutive number.

The names for variables and channels are derived from those of the graph's edges, if available. These names are constructed with the prefixes $c.$ or $v.$ for channels and variables, respectively. They are followed by the edge's name and the suffix *.cnr*, where *cnr* is a unique consecutive number.

The internal channels (interprocess communication on one PE) are declared at the beginning of the PE's main process, whereas all variable declarations are placed at the beginning of the process in which they are used.

Prior to translating a linear cluster of data flow graph nodes into a (sequential) OCCAM process, an execution order must be determined. This serialization is achieved by associating levels with each node. The nodes receiving only input data are labelled level one, each direct successor node level two, and so forth. If a node is reachable from the process inputs on several paths then its level is the maximum of the levels of all its predecessors. The exact execution order is determined by sorting the nodes according to their levels. If two nodes possess the same level, then their relative position in the linear node list of the parent graph is used for the decision.

The file "*target_language.txt*" contains for each graph node the precise OCCAM sequence for the translation. Instead of the variables there are dummy names. These variables are replaced by the real names before writing the sequence to the source code file.

For each process, the serialized graph nodes are translated into a sequence of OCCAM statements. The input and output operations are performed sequentially in the execution order that has been established during the simulation of the graph's execution. Therefore, no deadlocks are possible. All processes located on the same PE are placed for quasi-parallel execution. The exact order of execution has not been previously determined due to the high computational cost of simulating the exact execution of several tasks on a PE.

Since not all the information is gathered from the data flow graph in the order in which it is needed in the OCCAM source code, some sequences such as variable declarations are buffered in temporary files. Only one pass through the data flow graph is thus necessary to generate the corresponding OCCAM program.

<u>OCCGEN</u>(fgr, fpe):
(* translates the function graph fgr to an OCCAM program *)
(* fpe is the pointer to the first PE in the system *)

begin
    protocol_nr := 0, link_nr := 0
    define the protocols for the multiplexed and fused links in occam.all
    define all channels between the PEs in occam.all
    for all PEs **do**
        create processor definition in occam.all
        - number and type (T8) of processor

OCCAM code generation

        - map the channels to the physical links
        - head of all procedures placed on the PE
    end for all
    for all PEs ape **do**
        indent := 0
        create process header PE_(ape.no) in occam.src with indent
        indent := indent + 1
        if some links are fused or multiplexed **then**
            create the additional processes for fusion, multiplexing
                in the files source.txt, proccall.txt
        end if
        for all tasks tpr **do**
            define the process name pname
            <u>create process</u>(tpr, pname, call.txt, chan.txt, src.txt)
            append call.txt to proc_call.txt
            append chan.txt to channels.txt
            append src.txt to source.txt
        end for all
        copy channels.txt to occam.src with indentation indent
        copy source.txt to occam.src with indentation indent
        if there is more than one task or multiplexed/fused links **then**
            create code for parallel processes in occam.src
            indent := indent + 1
        else
            create code for sequence of processes in occam.src
        end if
        copy proc_call.txt to occam.src with indentation indent
        create process ending PE_(ape.no) in occam.src, indent := 0
    end for all
    check occam.src for length of line <= 80 characters, otherwise
    break up lines into smaller sections with correct indentation
end occgen

<u>create process</u>(proc, pna, pe, calltxt, chantxt, srctxt):
(* creates a process of the nodes in proc, located on pe *)
(* calltext contains the process head for calling it *)
(* the external channel declarations are written to chantext *)
(* the source code including local variables is returned in srctxt *)

begin
    indentation := 0
    create process header with name pna
    if the proc. receives data from or sends data to the main input **then**
        define the input or output variables in the process header

        end if
        write the process header to srctxt and calltxt
        define the internal channels in chantxt
        if library functions are called **then**
            write "#USE libraryname" to srctxt
        end if
        define all channels in srctxt and chantxt
        declare all variables in var.txt
        serialize_nodes(proc, slist)
        for all nodes nd of slist **do** in ascending order
            transform_node(nd, node.txt, variables.txt)
        end for all
        indentation := indentation + 1
        copy variables.txt to srctxt with indentation
        copy var.txt to srctxt with indentation
        create code for sequence of instructions in srctxt with indentation
        indentation := indentation + 1
        copy node.txt to srctxt with indentation
        create end of process in srctxt with indentation 0
end create_process

serialize_nodes(sproc, serlist):
(* imposes a serial execution order on the nodes of task sproc *)
(* serlist contains the nodes in the correct execution order *)

begin
    find nodes with only external inputs, set level := 1
    find nodes with only constant inputs, set level := 1
    for all successor nodes nd **do**
        determine the level of nd
    end for all
    sort the nodes in ascending order of the level, set node.execute
end serialize_nodes

transform_node(nd, ndtxt, vartxt):
(* ndtxt   :   source code text file representing the node *)
(* vartext :   variable declaration text file used for the node *)

    begin
        if there are external inputs or outputs **then**
            define the channels
            write the definitions to channels.txt
        end if
        case nd.class **of**

```
        Simple:   transform simple(nd, ndtxt, vartxt)
        Compound:   transform compound(nd, ndtxt, vartxt)
        Graph:   error(illegal node type)
    end case
end transform_node

transform simple(nod, nd.txt, var.txt):
begin
    read the node's code sequence from the file target_language.txt
    write it to the buffer srcbuf
    determine the names or the values of the input and output edges
    write the variable declarations to the file var.txt
    write the external input statements in the specified order to nd.txt
    copy the buffer srcbuf tond.txt with inserted variable names
    write the external output statements in the specified order to nd.txt
end transform_simple

transform compound(nod, nd.txt, var.txt):
begin
write the variable declarations to the file var.txt
write the external input statements in the specified order to nd.txt
case nod of
        Forall:   error(node should have been eliminated)

        Select:
            begin
                define the local input and output variables of the
                    alternative graphs in nd.txt
                create code for a sequence of instructions in nd.txt
                read the header sequence from target_language.txt
                write it to the buffer srcbuf
                copy srcbuf to nd.txt with inserted variable names
                    with indentation 1
                for all alternative graphs agr do
                    read case sequence from target_language.txt
                    write it to the buffer srcbuf
                    insert the alternative number
                    write it to nd.txt with indentation 2
                    create code for a sequence in nd.txt
                        with indentation 3
                    create process(agr, proc.txt)
                    copy proc.txt to nd.txt with indentation 4
                end for all
            end Select
```

TagCase: error(illegal node)

LoopA:
  begin
    define local input and output variables of the
      sub-graphs in nd.txt
    create code for a sequence of instructions in nd.txt
    create in nd.txt with indentation 1 the processes for the
      - initialization subgraph
      - body subgraph
      - test subgraph
    read code for while loop from target_language.txt
    write it to the buffer srcbuf
    copy the srcbuf to nd.txt with indentation 1 with
      inserted test variable
    create code for a sequence of instructions in nd.txt
      with indentation 2
    create in nd.txt with indentation 3 the processes for the
      - body subgraph
      - test subgraph
    create in nd.txt with indentation 1 the processes for the
      - returns subgraph
  end LoopA

LoopB:
  begin
    define local input and output variables of the
      sub-graphs in nd.txt
    create code for a sequence of instructions in nd.txt
    create in nd.txt with indentation 1 the processes for the
      - initialization subgraph
      - test subgraph
    read code for while loop from target_language.txt
    write it to the buffer srcbuf
    copy srcbuf to the file nd.txt with indentation 1 with
      inserted test variable
    create code for a sequence of instructions in nd.txt
      with indentation 2
    create in nd.txt with indentation 3 the processes for the
      - body subgraph
      - test subgraph
    create in nd.txt with indentation 1 the processes for the
      - returns subgraph
  end LoopB

Iter:
  begin
    define local input and all output var. of the body graph
      in nd.txt
    create code for a sequence of instructions in nd.txt
    write code for copying all input values to the output
      variables to nd.txt with indentation 1
    read code for while loop from target_language.txt
    write it to the buffer srcbuf
    copy srcbuf to the file nd.txt with indentation 1 with
      inserted test variable (= output 1)
    create code for a sequence of instructions in nd.txt
      with indentation 2
    <u>create_process</u>(initialization subgraph, proc.txt)
    copy proc.txt to nd.txt with indentation 3
    create code for copying the external outputs to the
      corresp. channels in nd.txt with indentation 1
  end Iter
end case
write the external output statements in the specified order to the file nd.txt
end transform_compound

Index

Bold entries in the index indicate the main treatment of a subject or the definition of the term.

Adam, T.L. 95
Adams, D.A. 12
adjacency matrix 72, 74
Aggarwal, J.K. 97, 99
aliasing 29
allocation 13, 81
    computational complexity of 105
    dynamic 82
    parameter 13, **101**, 113, 122, 125, 131, 132, 136, 137, 144, 170
    static 82, 84, **85**, 106
Arvind 13

backtracking 68
Baxter, J. 98
Bessel low-pass 31
bilinear transform 22, 31
bit reverse permutation 35
Blackman, R.B. 41
BLAS (Basic Linear Algebra Subroutines) 42, 43
block matrix computations **46**
Bokhari, S.H. 94
Bollinger, S.W. 97
Borrmann, L. 85, 92, 94, 99
Breadth-First Search (BFS) 67, 71
Briggs, F.A. 7
Burrus, C.S. 36
Bütler, B. 97

C 147

C++ 147
Campbell, M.L. 66
CDFG (Compressed Data Flow Graph) 66
Chandy, K.M. 95
channel **11**
    external **59**, 61, 72, 106, 173
    internal **59**, 61, 106, 173
    virtual **10**
Cholesky decomposition 47
cluster
    analysis 97, 98
    linear 146
clustering 99
    linear **97**, 99
    nonlinear **97**, 99
code generator 107
communication
    analysis 153
    blocking 69, 76
    cost 2, 59, 66, 97, 103, 132
        measurement 60
        model 59, 60, 85, 92, 106
    delay 8
    external 116
    interprocessor 13
    minimization 63
    network 85, 89, 96
        topology 90, 97, 101
    nonblocking 69
    overhead 61, 63, 64, 85, 90, 137, 143, 144
    resource utilization 91, 99, 103

182   Index

resources 145
time
   predictable 10
   volume analysis 54, 115
   volume matrix **86**
communication cost 64
computation time
   upper limit 26
computational efforts
   Adams-Basforth integration 29
   Cholesky decomposition 47
   fast Fourier transform 36, 40
   FIR filter 31
   GAXPY 44
   IIR filter 31
   inner product 43
   Kalman filter 25
   linear state feedback 24
   linear state feedback with estimator 24
   LR decomposition 48
   LU decomposition 47
   matrix inversion 47
   matrix multiplication 46
   matrix-vector multiplication 44
   outer product 43
   PID controller 23
   Runge-Kutta integration 28
   SAXPY 44
   scalar-vector multiplication 43
   SISO controller 23
   vector addition 43
   vector multiplication 43
control flow computer **6**
controller
   dynamic 19
   LQG 24
   LQG/LTR 24
   LQR 131, 143
   MIMO **23**
   nonlinear 113, 134
   optimal 24
   PID 19, **22**, 121
   robust 24
   robustness 25
   SISO 19, 21, **23**, 43
   state feedback 135
   state space 19, 23, 113, 131
convolution 20, 40
   discrete 40
Cooley, J.W. 32
correlation 20
   discrete 41, 43
critical path 71, 72, 95, 96
cross-correlation
   discrete 41
CSP (Communicating Sequential Processes) 12
Curtis, B.A. 96

D'Hollander, E.H. 97
data 10
data acquisition system 149, 150
data dependencies 12, 57
data flow
   approach 2
   computer **6**, 69
   graph 2, 10, **12**, 13, 53, 63, 65, 69, 75, 76, 101, 106, 117, 144, 145, 153, 166, 169, 174
      conversion to OCCAM 107
      cycles in a 69
      execution simulation 76
      principle 7
data token 69
data type
   basic 56
   compound 56
data-driven computations **6**
deadlock 71, 146, 174
   avoidance 68, 76
   implementational **68**
   structural **68**
demultiplexer 103, 107, 174
Dennis, J. 12
dependencies
   artificial 1
   true 2
Depth-First Search (DFS) 67
Devis, Y. 97
Dickins, J.R. 95
difference equation 21, 25
differential equation
   linear 26
   nonlinear 26
   nonstiff 27
   ordinary 26, 27
   ordinary nonlinear 27
   stiff 26
discrete Fourier transform (DFT) 19, 32
distance matrix **86**
distributed-memory computer 5
dot product **43**
Duhamel, P. 32
dynamic controller 19

dynamic process management 106
Dynamic Programming 94

eigenvalue/eigenvector decomposition 42
EPL (Experimental Programming Language) 11
Esser, R. 97
execution cost 64
execution cost measurements 62
execution time
    minimum 169
expanded graph 57, 115, 116, 122

fast Fourier transform 4, 20, 30, 32, 113, 138, 143, 145
    basic concept 36
    butterfly diagram 33
    Cooley-Tukey 35
    mixed-radix 36
    prime factor 32
    recursive formulation 33
    Tensor Product 38
    twiddle factor 32, **36**
    two-dimensional 32
Feng, T.Y. 5
Fernández-Baca, D. 95
filter
    analog 29
    anti-aliasing 29, 150, 152
    digital 113, 121
    discrete 30
    low-pass 29
    mixed design 29
    switched-capacitor 150
FIR filter 19, **30**, 43, 113, 121, 122, 123, 124, 125
    design **30**
Flynn, M.J. 5
Ford, L.R. 94
Fortran 1, 10, 42
Fosseen, J. 12
frequency sampling 31
Fulkerson, D.R. 94
function
    call overhead 58, 116, 137, 143
    expansion 58, 139, 143
function controlalgorithm 11, 53, **54**, 56, 114, 115

Gaudiot, J.-L. 67, 106
Gauss, C.F. 32

Gaussian elimination 47
GAXPY **44**
Giloi, W.K. 5
Gligor, V.D. 96
Good, I.J. 32
graceful degradation 85, 88
graph
    directed 12
    edge 12
    expansion 56, 68, 155
    node 12
group scheduling 96

Händler, W. 5
hard real-time systems 7
hardware constraints 13, 66, 122
Harvard computer architecture 6
HDFL (Hughes Data Flow Language) 67
hierarchical graph structure 54
HLFET list scheduling 97
Hoare, C.A.R. 12
Hockney, R.W. 7
Hollman, H. 32
Houstis, C.E. 97
Huang, J.P. 66
Hughes Data Flow Machine (HDFM) 8
Hwang, K. 7
hypercube 85, 96, 97

IF1 53, 67, 70, 74, 106, 156, 163
IF1 (Intermediate Form 1) 13
IF1 Display 54
IIR filter 19, 31, 43, 113, 121, 127, 128, 130, 131, 137
    design **31**
image processing 30, 32
impulse response 30
initial value problem 26
inner product **43**
integration
    Adams-Bashforth 27, **28**
    Euler 27
    Heun 27
    multistep method 27, **28**
    one-step method **27**
    predictor-corrector 27
    Runge-Kutta 27, 113, 137, 143
interconnection
    interprocessor 4
    network 9, 59
interference costs 94

Intermediate Form 1 (IF1) 53
interprocessor communication 87, 93
    cost 97
intertask communication 88, 90, 97
    cost 88
intrinsic function 58

Jesshope, C.R. 7

Kalman filter 19, **24**, 47, 90
Kalman, R.E. 24
Kasahara, H. 96
Kathail, V. 13
Kronecker product 38

LAPACK (Linear Algebra PACKage) 46
Lee, L.-T. 67, 106
Lee, S.Y. 97, 99
Length of Longest Output Path (LLOP) 71
Linderman, J. 12
linear equation system 20, 41, **46**, 47
linear speed-up 9
linear state feedback 23
link 92, 97, 101, 107
    bidirectional 60, 91, 113
    external 166
    fusion 9, **102**, 103, 144, 170
    matrix 90, 92
    multiplexing 9, 64, **102**, 103, 144, 170
    serial 9, 59, 89, 99, 150, 169
LINPACK 43
list scheduling 93, 95, 96
Lo, S.P. 96
Lo, V.M. 94
load balancing 59, 87, 89, 104
Löffler, C. 96
longest path 96
LR decomposition 47
LU decomposition 47

Madisetti, V.K. 96
mapping
    optimum 13
matrix
    computations 20
    exponential 42
    inversion 20, 47
    multiplication 46
matrix-vector multiplication **44**
Mattmann, R. 97
May, M.D. 11

message passing 7, 69, 85, 99, 147
message-passing architecture 2, 99
message routing 9, 89, 144
Midkiff, S.F. 97
MIMD **5**
MIMO controller **23**
min-cut problem 65, 66
minimization problem 72
MISD **5**
module graph 66
monitor 64
multicomputer **5**, 13
    architecture classification **5**
    general 85
    heterogeneous **85**
    homogeneous **6**, **85**, 88, 90
    inhomogeneous **6**, 92
    message-passing **6**
multiplexer 103, 107, 174
multiprocessor system 46, 143
multi-rate signal processing 143

Narita, S. 96
node
    bottleneck **68**
    clustering 59, 63
    compound 53, 107
    cost table 106, 156, 160
    execution cost 13, **59**, 61, 153
    Forall 53, 54, 56, **57**, 115, 155, 156
    simple 53
numerical integration 19, 26
    algorithms **26**
    error 27
    explicit 26
    implicit 26
    stability 26

observer 23, 113
OCCAM 10, **11**, 14, 106, 107, 114, 119, 147, 149, 174
    code sequence for a graph node 107
OCCAM91 12, 146
optimum task assignment 91
outer product **43**
oversampling 29

parallel computer architecture 7
parallelism
    fine grain 64
    inherent 11, 53

main sources of 56
potential 20
preservation 63, 72, 73, 75, 144, 145
upper bound of 9
parallelizing computations 42
Partitioned Data Flow Graph (PDFG) 67
partitioning 13, 53, 63, 72, 74, 77, 117, 122, 140, 153, 166
aims of 63, 65
compile-time scheduling approach **67**
computational complexity of 77
efficiency 66
heuristic approach 66
macro-data flow approach **67**
optimum 65
Pascal 10, 153
Patel, J.H. 98
PE load 122
PE utilization 72, 118, 124, 131, 144
Pease, M.C. 37, 38
permutation matrix 37
Petri Net 69, 97
phase response
linear 30, 31
Pingali, K. 13
pipelining 13
pole placement 23, 31
POSC (Partitioning and Optimizing SISAL Compiler) 68
prediction estimator **24**
problem mapping
optimum 9
process **11**, 14
processor
load 82
utilization 82
program execution time 9, 13, 72, 91, 92, 93, 113, 116
minimum 8
partitioned 122
serial 122
Program Structure Graph (PSG) 67
PSR 116, 122, 123, 124, 125, 128, 130, 131, 132, 136, 137, 143, 144
PSR (parallel-to-serial execution time ratio) **113**

real-time
application 2, 113
computations 26
control 8

data processing 149
signal processing 106, 143
real-time system 7, 8, 82, 92
hard **82**
soft **82**
recursive function call 11
regular iterative algorithms (RIA) 20
Remez exchange method 30
rendezvous time 76
resource
capacity matrix **86**
demand matrix **86**
utilization 87
response time **81**
Runge-Kutta-Fehlberg algorithm 26

sampling rate 150, 152
Sarkar, V. 65, 67, 68
SAXPY **44**, 45
scalar product **43**, 113
schedule
non-periodic **82**, 89
optimum 94
periodic **82**, 89, 96
scheduler 145
run-time 3, 7
scheduling 81
cost 8, 64, 65, 67
dynamic 81
dynamic pre-emptive 7
greedy 96
incremental 96
round-robin 83
static 7
Schibli, P. 106
semaphore 64
serialization of the computations 64
shared-memory computer 5
Shen, C.-C. 94
Shield, D.T. 106
signal 64
Signal Processor
TMS320C40 146
signal transform 30
SIMD **5**
simulated annealing 94, **96**
Sinclair, J.B. 95
Single Assignment principle **10**
singular value decomposition 42
SISAL 10, 13, 53, 67, 68, 106, 113, 115, 146, 153

SISD **5**
SISO controller 19, 21, **23**, 43
Skillicorn, D.B. 5
source code generation 106
spectrum estimation 20, 30, 41
speech processing 30, 32
Stasinski, R. 32
state
    feedback 19
state space
    controller 19
    representation **20**, 25
Steele, C.S. 96
Stone, H.S. 85, 94
Strassen, V. 47
substitution
    backward 47, 48
    forward 47, 48
super-cluster 146
synchronization 2
    overhead 63
system
    causal 20, **23**
    discrete 20
    linear 19, **20**, 25
    nonlinear 19, **21**, 25, 27
    only time-invariant **21**
    resources **81**
    utilization 81, 89, 91

tagged token 13
target language 106, 107
task 13, 63, 106
    administration cost 63
    administration overhead 146
    allocation 153, 169
    allocation problem 93, 99
    allocation procedure 99
    assignment 81
        matrix **85**
        problem 86, 92
    average number per PE 101, 103
    clustering 72, 94
    coresidence 88, 91
    execution cost 88, 93
    execution time 8
    granularity 144

graph 94, 95, 97
heuristic allocation 93
merging 67, 72
modified assignment problem 90
parallelism matrix 90
pipelined execution 89
re-serialization 146
scheduling 66
serially dependent 99
static allocation 92
synchronization 64
turnaround time 7
Taylor, R.J.B. 11
Temperton, C. 32, 38
Thaler, M. 96
time delay 30
transfer function 25
    continuous **20**, 22
    discrete **20**, 22
transitive closure 67, 68, 72, 75
Transputer 2, 8, 14, 106, 145
    T222 150
    T800 7, 9, 60, 76, 89, 116, 146, 149, 150, 152, 163
    T9000 10, 12, 146
triangular matrix 47
Tsai, W.H. 94
Tukey, J.W. 32, 41
twiddle factor 36

VAL (Value-oriented Algorithmic Language) 10, 67
vector
    addition **43**
    computer 41
    multiplication **43**
volume of communication
    analysis **56**
von Neumann computer architecture 6

Warshall's algorithm, 75
wavefront analysis 96
Welch, P.D. 41
Whitby-Strevens, C. 11
Winograd, S. 32
Wüst, U. 106

Zverev, A.I. 31